SEEDBOMBS
GOING WILD
with FLOWERS

Josie Jeffery

SEEDBOMBS
GOING WILD
with FLOWERS

Leaping Hare Press

First published in the UK in 2011 by

Leaping Hare Press

210 High Street
Lewes
East Sussex BN7 2NS
United Kingdom
www.leapingharepress.co.uk

British Library Cataloguing-in-Publication Data
A catalogue record for this book is available from
the British Library

ISBN: 978-1-907332-55-5

This book was conceived, designed,
and produced by

Leaping Hare Press

CREATIVE DIRECTOR Peter Bridgewater
PUBLISHER Jason Hook
COMMISSIONING EDITOR Monica Perdoni
ART DIRECTOR Wayne Blades
SENIOR EDITOR Polita Anderson
DESIGNER Clare Barber
PHOTOGRAPHER Neal Grundy
ILLUSTRATOR Andrew Farmer

Printed in China

10 9 8 7 6 5 4 3 2 1

Contents

Josie and her Roots

I had a creative, musical and unusual upbringing. My family home was made from wood and metal and had wheels! I lived in a bus with my family and was home-educated. I'm the second of five kids – the others are Amy, Arran, Holly and Rose.

During the late 1970s and 80s we travelled around Europe in a convoy or on our own and busked as a family band to make a living. We went to festivals like Stonehenge and Glastonbury, squatted land and travelled with circuses.

We used to rest from our travels at a commune in Suffolk called Brick Kiln Farm. We grew food and reared chickens and pigs.

Brick Kiln Farm ran a charity called Green Deserts and grew trees in nurseries on the farm. I remember running around barefoot at Rougham Tree Fair, a charity fundraiser organized by Green Deserts, getting up to all sorts of mischief!

Green Deserts' inspiring ethos was to refertilize desert wastelands using natural energy technology and organic plant husbandry.

My family settled in Wales in 1990 at a friend's farm, and in 1991 we bought some Welsh land – somewhere to anchor our roots.

Arran, Holly and Rose went to school and Amy and I began college. I studied art and met Steve, the father of my three sons – Tyrone, Isaac and George.

In 2003 we moved to Brighton, where – inspired by my love of watching things grow – I studied horticulture, then garden design. I began designing gardens that were slightly unruly and anarchistic. I veered towards recycled materials, wayward plants and graffiti murals.

During college I had a job at a local plant nursery and also helped run children's gardening workshops at the Museum of Garden History in London.

I heard the word 'seedbomb' on the radio and a 'ping' moment happened – next thing I know we were doing seedbomb workshops in the museum, and I went on to do workshops at festivals and fairs, art projects, schools and for Brighton and Hove food partnership.

I set up my business, seedfreedom.net, and am now writing this book!

AN INTRO TO SEEDBOMBING

WHAT IS A SEEDBOMB?
The History of Seedbombing

When I tell people I make seedbombs, they look puzzled and ask, 'What is a seedbomb?'. They think they are edible (some fancy new superfood) or a cosmetic product. Rarely do people think they are horticultural. I smile and begin a well-rehearsed explanation.

Firstly, they are **NOT EXPLOSIVE OR EDIBLE!**

A seedbomb is a little ball made up of a combination of compost, clay and seeds.

'WHAT IS IT FOR?'

The compost and clay act as a carrier for the seeds so they can be launched over walls or fences and into inaccessible areas such as wasteland or railways.

'BUT WHAT IS THE POINT? WHY CAN'T YOU JUST THROW SEEDS LOOSE?'

Most seeds are very light and there is risk of them being blown away by the wind, making them unsuitable for launching long distances.

TWO ROLL YOUR BALL ...

FOUR WATCH IT FLOURISH!

'HOW DO I MAKE THEM?'

There are various ways of making seedbombs. You need to find a carrier for the seeds. My method uses natural ingredients – compost and clay. The compost offers nutrients for the seeds to germinate and grow strong during their infancy and the clay binds the seedbomb, making it hard enough not to break when it hits the ground.

'HOW DO THEY WORK?'

After about 3 weeks the first seedlings work their way through the seedbomb and root into the ground below. The seedlings will then grow into mature plants and face whatever conditions Mother Nature has in store for them. As they grow, more seeds germinate and the seedbomb begins to dissolve. This can take days, weeks or months – it depends on the quantity of rainfall.

Seeds will remain dormant until their environmental needs are met with these factors: water, correct temperature and a good position to grow in.

ONE MIX YOUR SEEDS ...

THREE
LAUNCH YOUR
SEEDBOMB ...

There is a sense of unpredictability with seedbombing. Its random nature is what attracts people, the magic of waiting to see if this strange little ball will grow ... if it actually works.

THE BEAUTY OF SEEDBOMBING

Here you have in the palm of your hand a little revolution, something that can change the face of the earth, something that contains the early stages of a field of wild flowers, edible crops or a herb garden.

You can use seeds of one plant for your seedbombs or combinations of compatible seeds. This is called 'companion planting', where you use plants that grow well together and assist each other in a number of ways, such as pollination, deterring pests and soil conditioning. With a little help from Mother Nature, something as small as a seedbomb has the potential to improve the natural structure of an area in one fell swoop.

Small but powerful, they have the potential to improve an area.

SENSIBLE SEEDBOMBING

A seedbomb is a little ball of life and comes with a responsibility to choose your plants and be used in the correct way.

You have to consider not only the environment where you choose to launch your seedbomb, but also the welfare of the plant. It is your job as a gardener/seedbomber to make sure the seeds get a good chance of germinating and have a good probability of reaching infancy and – even better – maturing into plants that flower and fruit and connect the ends of the circle of growth.

LAUNCH CHECKLIST

✔ Launch them at the right time of year so the right temperature is achieved

✔ Check the weather forecast – rain is good!

✔ Right plant, right place

AND AWAY WE GO!

The seeds may germinate next week or next spring or not at all.

But we still want to try it!

Launch your seedbombs into desolate areas or wastelands and watch as nature begins to reclaim.

WHY ARE THEY USED?

Seedbombing is another form of seed dispersal, a human intervention into what is already happening in nature anyway. It is an efficient way of deliberately dispersing seeds, but trying to work in harmony with nature, too, by being considerate of wildlife and natural habitats.

Seedbombs are seeds wrapped up in a blanket of earth, which acts as a carrier for the seeds and enables them to be launched in areas that are physically challenging to access, like fenced-off wasteland or motorway banks. Because of their size and strength they can simply be thrown over the fences or out of car, train or bus windows. They provide the accuracy needed to get the plants to where you want them to grow. It does help if your aim is good though!

The risk of damage to the seeds is minimized because the blanket of earth offers protection from harsh weather conditions and seedeaters such as mice and birds.

Apart from the fact that they are really good fun to make, they actually work and it's so exciting when you see the first shoot come through!

ABOVE/BELOW SEEDBOMBS ALLOW THE ACCURACY NEEDED FOR DISTANCE SOWING

Like leaving a trail of breadcrumbs or petals for a loved one, seedbombing is a kind of footprint, a marker for your journeys.

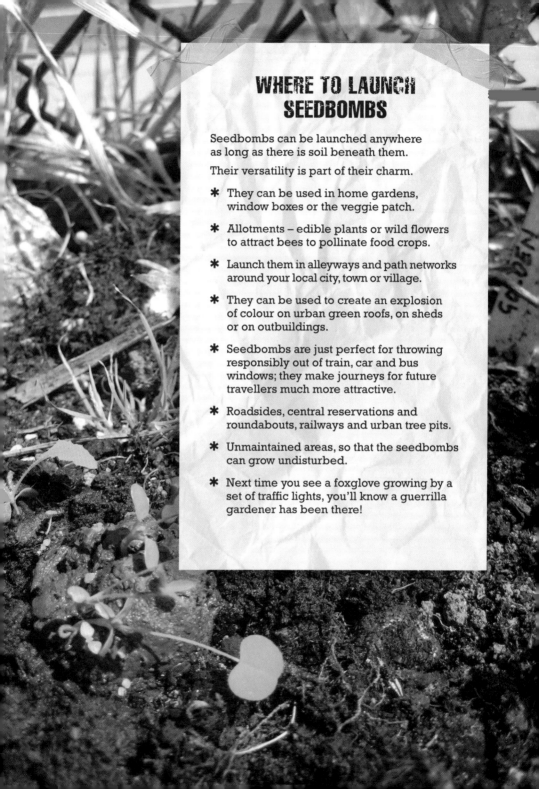

WHERE TO LAUNCH SEEDBOMBS

Seedbombs can be launched anywhere as long as there is soil beneath them.

Their versatility is part of their charm.

* They can be used in home gardens, window boxes or the veggie patch.

* Allotments – edible plants or wild flowers to attract bees to pollinate food crops.

* Launch them in alleyways and path networks around your local city, town or village.

* They can be used to create an explosion of colour on urban green roofs, on sheds or on outbuildings.

* Seedbombs are just perfect for throwing responsibly out of train, car and bus windows; they make journeys for future travellers much more attractive.

* Roadsides, central reservations and roundabouts, railways and urban tree pits.

* Unmaintained areas, so that the seedbombs can grow undisturbed.

* Next time you see a foxglove growing by a set of traffic lights, you'll know a guerrilla gardener has been there!

A SEEDBOMB HISTORY

If something has been used effectively throughout time, it is given validity; you know it has been tried, tested and perfected. When it comes to seedbombs, one man has done just that. Masanobu Fukuoka incorporated his ancestral gardening techniques into his own farming methods and, in so doing, started a revolution.

Seedbombs are an ancient Japanese practice called *Tsuchi Dango*, meaning 'Earth Dumpling' (because they are made from earth). They were reintroduced in 1938 by the Japanese microbiologist/ farmer and philosopher Masanobu Fukuoka (1913–2008), author of *The One Straw Revolution*.

Fukuoka led the way into the world of sustainable agriculture by initiating 'natural farming'. His methods were simple and produced no pollution. His technique used no machines or chemicals and almost no weeding.

Seedbombing was part of Fukuoka's annual farming regime. He believed that Mother Nature takes care of the seeds we sow and decides which crops to provide us with, like a process of natural selection, because ultimately nature decides what will grow and when germination will occur, be that in 7 days or several seasons away.

ABOVE FUKUOKA'S APPROACH TO SUSTAINABLE AGRICULTURE HAS INSPIRED GENERATIONS.

RIGHT FUKUOKA'S METHODS WERE SIMPLE AND NON-INVASIVE.

Fukuoka grew vegetables like wild plants – he called it 'semi wild'. He seedbombed on riverbanks, roadsides and wasteland and allowed them to 'grow up' with the weeds. He believed that vegetables grown in this way – including Japanese radish, carrots, burdock, onions and turnips – are stronger than most people think.

He'd add clover to his vegetable mixes because it acted as a living mulch and conditioned the soil.

PROJECTS

Where land has been intensively farmed by modern methods, the natural fertility is destroyed. Fukuoka used clover seedbombs to rehabilitate natural fertility on dead land so that food crops could be grown successfully again.

In 1998, hundreds of locals in Arnissa, Greece – families, schoolchildren, ministers, farmers, journalists – were inspired by Fukuoka's work to partake in launching tons of seedbombs over 10,000 hectares of desolate land damaged by human activity. The seeds were donated by the Ministry of Agriculture and the National Institute of Agrarian Research.

GREENING DESERTS WITH SEEDBOMBS

Fukuoka encouraged people to collect seeds and instigated a movement for desert-greening with seedbombs. He successfully 'greened up' land all over the world, including Greece, India and the Americas.

Through his works, land has come 'alive' again, with plants and wildlife and food for the people.

Barley crop with clover, weeds and whatever else nature has in store!

ABOVE FUKUOKA USED NO MACHINES OR CHEMICALS IN HIS FARMING METHODS.

Seedbombs are a small universe in themselves.
MASANOBU FUKUOKA

SEEDBOMB BENEFITS

Seedbombs are full of potential wrapped up in a pocket-sized ball of mud! They can make ugly, forgotten land beautiful and useful again; restore plant and wildlife populations; nourish and feed the soil, people and animals; bring communities together, educate and – importantly – bring joy.

NATURE

As seen with Fukuoka's work, there are countless benefits to using the seedbomb method for planting. They are successfully used to grow food crops and meadowland and replant areas that have suffered from drought or forest fire damage, as well as being suitable for small-scale gardens and allotments.

Seedbombs help to create vegetation in areas where it is absent or sparse due to the land being neglected. They present the prospect of attracting wildlife such as bees, birds and butterflies into our urban environment, making them available to pollinate our city flowers and food crops and help the production of fruit and seed.

Seedbombs can be used to repopulate an area with diminishing numbers of wild flowers, which in turn will attract wildlife that may, too, be diminishing.

URBAN ENVIRONMENT

On an aesthetic level, the plants provide colour and character to otherwise unattractive sites that are usually broken into and could be littered with rubbish and the classic shopping trolleys and broken glass. Areas such as this are in danger of fostering feelings of apathy, an *'Oh, what's the point, the place is a mess anyway'* attitude.

RIGHT YOUR SEEDBOMB COULD ATTRACT NEW WILDLIFE TO PREVIOUSLY DESERTED AREAS.

ABOVE AN INACCESSIBLE EYESORE CAN BE TRANSFORMED INTO A THING OF BEAUTY.

Some sites stay undeveloped for a number of years and become part of the personality of the area; numerous people will walk past these places every day and not even notice the potential. Desolate spaces may be used by people acting undesirably and are often a location for fly-tipping. Unfortunately, this kind of activity could be a danger to the public because the fences get pulled down, leaving these spaces open to adventuring children.

Often no one wants to claim responsibility and implement a clear-up because ownership of the land could be a grey area, while some sites are pending construction. Guerrilla gardeners target areas like these as potential sites to glam up with donated plants and recycled, reclaimed and reused objects.

As well as being fun, seedbombing can bring people together with an aim to beautify an area.

COMMUNITY

Guerrilla gardens are an open forum for people to make a positive contribution to ugly areas within their community that may have a negative impact not only aesthetically but demographically too. Abandoned wasteland can encourage antisocial behaviour and crime, whereas a sense of ownership, however temporary it may be, will encourage local communities to work together on improving their neighbourhood.

Some guerrilla gardeners work stealthily and under the cover of darkness, wielding spades and plants and a torch! It's a bit tricky and is often where seedbombs prove their worth. The anonymity of seedbombing is part of the appeal; a five-second action is less demanding than physically planting something, therefore making it easier for reluctant guerrilla gardeners to join in – no fence-scaling needed here!

Seedbombs are fun and easy to make and can be an activity that appeals to all age groups and walks of life, from farmer to amateur grower. Community gardens offer a space to have group gardening days and local events, like seed collecting, swapping and the making and launching of seedbombs – and that's gotta be good, right?

Some guerrilla gardeners will dedicate their nights to brightening up their area – a lovely morning surprise!

OURSELVES

Making seedbombs is such a relaxing activity to do on your own or with the family. Kids love the whole process as it involves design, dexterity, consideration, communication, maths, botany and a big dollop of patience. And the elation and pride felt when you launch the seedbomb and when you see the seedlings appearing and maturing into flowering plants is priceless.

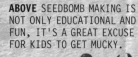

ABOVE SEEDBOMB MAKING IS NOT ONLY EDUCATIONAL AND FUN, IT'S A GREAT EXCUSE FOR KIDS TO GET MUCKY.

CONTEMPORARY SEEDBOMBS

Salad Burnet

THE INTENTIONAL SEEDBOMB

Historically, seedbombs were made with natural materials, but the guerrilla gardening movement has highlighted their usefulness and people have started developing their own vessels for carrying seeds. As environmental awareness spreads, people are beginning to take action through the medium of plants.

Seedbombing is an effective way of causing real change; they can be made very easily, stored efficiently, and some have a long shelf life. And there are endless ways to launch them.

AERIAL SEEDBOMBING

There are some very sophisticated modern seedbomb designs resembling NASA-type capsules or military missiles. These are designed for long-distance launching and used for purposes such as aerial reforestation to combat or replenish areas that have suffered desertification.

The capsules may contain seeds or seedlings, soil, nutrients, fertilizer and other materials that will help the plants' survival. The biodegradable casing protects the plants on impact with the ground and then disintegrates to let the roots emerge.

The skies could be filled with military planes dropping seedbombs instead of missiles.

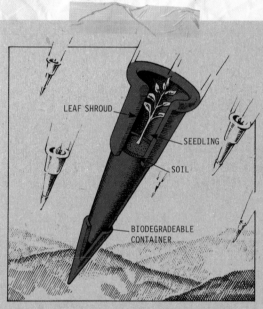

LEAF SHROUD

SEEDLING

SOIL

BIODEGRADEABLE CONTAINER

FLYING IN SEEDS

After forest fires in the 1930s, planes were used to distribute seeds over inaccessible mountains in Honolulu.

Some seedbombs have been designed with the military in mind for the distribution, not only for the irony but because they are professionals at launching bombs. They have the people power, the skills and the technology for tracking weather conditions and wind speed to ensure the best possible accuracy.

One large plane could drop up to 100,000 seeds in one flight – around a million trees in just one day. Even if a certain percentage of those dropped is unsuccessful that is still a lot of trees!

In 1997, Moshe Alamaro, an Israeli former aeronautical engineer, is known to have been working on a project to drop one-year-old tree seedlings in biodegradable open-topped cones from aircraft in order to reach previously inaccessible areas such as war-torn battlefields, deserts and slopes.

Travelling at 200mph, the cones would embed themselves in the soil and decompose before the seedlings took root.

There is little information about the success of projects like this and some are still in the design process, but even the fact that people are thinking of using seedbombs as an effective replanting process on such a grand scale validates just how ingenuous they actually are.

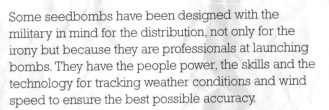

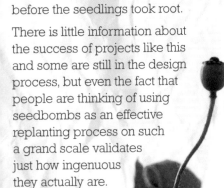

RAILWAY SEEDBOMBING

I met a train driver who had decided to enhance his route by throwing foxglove seeds out of the window. Now, thanks to him, year in year out there will be a bank of purple foxgloves to brighten up the monotony of a train journey! I like the idea that some people will sleep through them and others will spot them out of the corner of their eye, some will enjoy a moment with them before they whizz by – either way, they will be there, lined up like passengers on a platform, waiting for their bee visitors.

WET BALLOON BOMBS

Some people use balloons, like the old favourite water bombs of our childhood days. They explode on impact with the ground, but though they act as a suitable carrier for the seeds, they may not be particularly accurate and will leave the seeds vulnerable to attack by weather and seedeaters.

Place a funnel in the neck of the balloon and filter in the seeds and earth, then add water and fertilizer to the mix.

The balloon bomb is good to go!

It needs to be made and launched within the same day so as to avoid rotting within the bomb.

DRY BALLOON BOMBS

Balloon bombs can be made without liquid and simply stuffed with seeds and earth and filled with helium; these are lots of fun to launch into the atmosphere – eventually they'll come down, though accuracy is highly variable.

REMEMBER ...

It's important to use biodegradable balloons to prevent the remnants from polluting the environment.

These can take 4–6 months to decompose, which might be a concern if they are floating about in the atmosphere, or if they fall but do not burst or are impenetrable. When will the seeds germinate?

Be sure to choose your launch time carefully!

Spring launching will still provide some rainfall – this will help land and degrade the balloon, as well as giving the seeds enough time to germinate when the right conditions occur.

Autumn launching will give the balloon six months to land, degrade and germinate in time for spring.

Please note that it is a very variable and unpredictable method and at the end of the day Mother Nature will always decide.

As I looked up all I could see were balloons just waiting to land and burst with blooms.

NATURAL SEEDBOMBS

The most environmentally sound way of making seedbombs is to use dirt.

Dirt, lovely dirt! Whether it's from your garden or a garden centre, it contains everything needed for a plant to grow into a healthy seedling.

Homemade compost, humus, green manure and tea and coffee waste, leaf mould and chicken manure can be used, as well as natural binding materials like waste paper pulp and clay.

The idea of a seedbomb is that it is aerodynamic and able to travel a distance when thrown, so it naturally tends to be ball-shaped. Seedbombs, seedballs, earth dumplings, clay bombs: they are generally always the same form – small, round and pocket-sized – but you can form them into any shape you like or put them into moulds. Obviously this may counterbalance the aerodynamics — but you could simply lay them on the earth if you choose.

Interesting 'ironic' design ideas include gun-shaped seedbombs that

Compost, manure or di... a natural choice for bindin... your seedbomb.

BELOW CLEVERLY DESIGNED SEED GRENADES 'EXPLODE' ON HITTING THE GROUND.

ABOVE SEEDBOMBS CAN BE ANY SHAPE OR SIZE YOU CHOOSE — JUST BEAR IN MIND THE LAUNCH.

ABOVE/BELOW INNOVATIVE SEEDBOMB BANGLES ARE A FASHIONABLE METHOD OF CARRYING YOUR SEEDS.

sprout to give the appearance of a green gun and fired terracotta pots in the shape of grenades that release an explosion of compost and seeds.

Or you can make heart shapes and even flower shapes – just make sure there is enough depth for the plant to grow.

PAPER BAG BOMBS

Fill brown paper bags with a mixture of compost and seeds, leaving enough room to twist the top of the bag and tie a knot in it.

If the soil is moist, launch them on the same day.

If the soil is bone dry, they may keep for a number of weeks in a cool, dry and dark place, but do check that they are not mouldy before you launch them.

EGG BOMBS

Eggs can be blown and carefully filled with very dry powdery earth and seeds. They are quite fiddly to make but can be effective and are quite exciting when they smash onto the ground!

You can decorate them or write messages as a tribute, or to mark a celebration.

SEEDBOMB ACCESSORIES

I've seen some pretty cool seed accessories such as little pots full of seeds that can be worn on a chain as a pendant, a key ring, a charm bracelet or as earrings, and bangle bombs made from transparent tubes that can be refilled time and again with seeds and sprinkled when and where you choose. As well as distributing seeds, these accessory vessels can be used to collect seeds on your travels.

Whatever you use to make your seedbombs, think about the impact it will have on the environment, not just on the ground when it lands.

THE ACCIDENTAL SEEDBOMB

I have heard stories about people who have made their own seedbombs without realizing it. There is a world of anonymous dispersal happening without us even noticing and we wonder, 'How did that nigella end up in my garden? I know I didn't plant it...'

TOM' BOMBS

You're having a picnic in the garden and a tomato falls on the ground. You pick it up and throw it in the flower bed without even thinking about it and a few weeks later notice a plant growing! Just let it do its thing and you'll have tomatoes just in time for the final summer barbecue. Delicious!

BOOT BOMBS

You can pick up seeds on your boots simply by walking around outdoors, in your own garden, at the local park, out in the fields, in the woods … Your boots build up a fair bit of mud, which you knock off on the garden wall. The mud lands in the flower bed – and before you know it you have some guests in the garden.

ROADSIDE SEEDBOMBING

On outings, the tyre treads of your car pick up mud and dust; it spends most of its time parked in the driveway, next to a flower border. You pull up at some traffic lights and a clump of mud falls out of the tread. You drive off again, the wind blows the clump onto the verge, and when the next rain falls a tiny seed begins to germinate. Six weeks later you stop at those same traffic lights and think to yourself, *'How odd — aquilegias growing by the side of the road!'*

FEEDER BOMBS

Leftover seeds in bird feeders that have fallen from the tree at the bottom of the garden will sprout into a wonderful array of grasses.

WHAT IS GUERRILLA GARDENING?

HISTORICAL KEY POINTS

Guerrilla gardening is seen as a relatively new movement but people have been gardening this way for centuries – it's just that no one put a name to it. Here are some historical key figures and events that 'sowed the seed' for this growing (and expanding!) movement.

THE DIGGERS/ TRUE LEVELLERS

In 1649, a time of great social unrest in England, a group called the 'True Levellers' was formed, led by Gerrard Winstanley. Later known as 'The Diggers', they were Protestant Christian agrarian radical communists who believed there is an ecological interrelationship between humans and nature and that the inherent links between people and their surroundings need to be recognized.

> True freedom lies where a man receives his nourishment and preservation, and that is in the use of the earth.

GERRARD WINSTANLEY

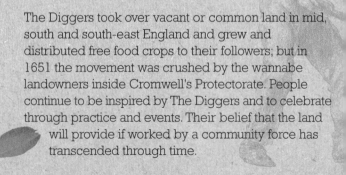

The Diggers took over vacant or common land in mid, south and south-east England and grew and distributed free food crops to their followers; but in 1651 the movement was crushed by the wannabe landowners inside Cromwell's Protectorate. People continue to be inspired by The Diggers and to celebrate through practice and events. Their belief that the land will provide if worked by a community force has transcended through time.

WHEN THE PHRASE WAS COINED

The revolutionary city of New York – the 'Big Apple' – bore the fruit and sowed the seed of a group of people who have inspired the guerrilla gardeners of today. In 1973 the first community garden in New York was developed and was eventually given permission to be rented in 1974 as the Bowery Houston Community Farm and Garden. The group was called the 'Green Guerrillas' and some say that is where the 'Guerrilla Gardener' phrase was coined.

In 2002 the City of New York and the NYS Attorney General approved the garden in its entirety and ordered that its renovation be continued.

JOHNNY APPLESEED DAY MARCH 11

This celebrates the work of American pioneer John Chapman (1774–1845), nicknamed 'Johnny Appleseed' because he used to collect apple seeds and spread them randomly everywhere he went. Orchards of over 15,000 trees sprung up all over Ohio, Illinois and Indiana due to this man! His work has made him a celebrated figure in America, through songs, comics and even a Disney animation, *The Legend of Johnny Appleseed.*

POLITICAL PRACTICE

'Guerrilla' is the Spanish word meaning a 'small army' that fights a stronger force. Perhaps the word 'seedbomb' has come about from its association with 'guerrilla', but if they are used responsibly they are certainly not belligerent bombs. Seedbombing is like a subdivision in a small army that could be called 'the stealth sowing division'!

Guerrilla gardening can swiftly bring an urban environment closer to nature ...

It's important that guerrilla gardeners only choose areas that are clearly vacant, and plant non-invasive or native plants.

A lot of guerrilla gardens are created on land that is to be developed with either housing or supermarkets. Unfortunately, impending constructions mean that some gardens have only a short life span. Fortunately, though, some gardens have successfully won over the community, landowners and the council and been granted tenancy.

BELOW/RIGHT
... AND ENHANCE A SENSE OF COMMUNITY.

The **POSITIVE** effects these 'pop up gardens' have on the community are priceless. New relationships are formed with people, with plants and with the environment. It only takes a handful of people to make a difference to an area; more hands make light work.

The **NEGATIVE** effects are that some people may not understand the ethos behind guerrilla gardening and instead see it as vandalism and crime. It may make some members of a community feel vulnerable, perhaps as if the neighbourhood is changing; they may worry that people/ strangers are attracted to the garden and may be using it for antisocial behaviour. So it is important to communicate with the neighbourhood and keep the activities in the garden positive – this can easily be done through putting up posters and flyers and inviting people to join in with the gardening or event days.

ABOVE ENSURE LAND IS CLEARLY VACANT BEFORE GARDENING, OTHERWISE IT COULD BE SEEN AS VANDALISM.

A group of people wielding spades, forks and plants is bound to raise eyebrows and get people talking and naturally they'll want to join in.

Guerrilla gardens don't have to be full of plants that are just for ornamental purposes. The home-grown-food revolution is upon us and people are taking matters into their own hands to obtain certain levels of 'food security' by growing their own food locally and setting up community projects to educate people on how to grow their own food.

GUERRILLA GARDENING
Groups and Individuals

Let's have a look at just a few examples of the many guerrilla gardening groups and individuals in the UK – but we'll start with New York's Green Guerrillas, because they were such an important catalyst to the movement and all research leads us back to them as being at its roots.

GREEN GUERRILLAS

This group was set up in New York in 1973 by an artist, Liz Christy, who assembled some of her friends and neighbours to clear out a vacant lot, where they created a vibrant community garden – the start of the New York guerrilla gardening movement.

The Green Guerrillas have beautified many desolate spots around the city and still operate today. Their mission is to bring the people together to create and educate through community gardens.

THE HUMAN SHRUB

One man (or woman) on a mission to make Colchester beautiful again!

In 2009 the human shrub took action against the council, who had drastically slashed their plants budget. Under the cover of moss and leaves he planted flowers and protested to the council, who reversed their plans to concrete over one in five roadside flower beds.

'THREE CHEERS FOR THE HUMAN SHRUB!'

ABOVE LIZ CHRISTY'S VISION OF A COMMUNITY GARDEN HAS ENCOURAGED GENERATIONS TO PITCH IN.

GLASGOW GUERRILLA GARDENERS

'Resistance is Fertile'

In 2008 volunteers grouped together with the collective aim of beautifying Glasgow.

Organized events are publicized on their website (www.glasgowguerrillagardening.org.uk) and social network sites. Their projects include planting 10,000 bulbs, seedbombing dreary walkways and planting up wild flowers to attract bees in the city centre.

LONDON GUERRILLA GARDENERS

In 2004 guerrillagardening.org was set up to document events that happened in London, network with likeminded people, dig in, seek advice and even get a troop number!

BRIGHTON GUERRILLA GARDENERS

In May 2009 Brighton's first guerrilla garden was built by community troops. Wielding plants, forks and wheelbarrows, they transformed an abandoned petrol station from a bleak wasteland into the Lewes Road Community Garden. It was open for one blissful year before the 'heavies' came in and forced out all living things. What was once a place for family picnics, relaxing and bonding with people and plants has been claimed back by industry to become another block of consumer concrete.

Guerrilla gardening can promote respect for the global environment.

ABOVE GUERRILLA GARDENERS FEEL THAT WASTELAND SHOULD BE TURNED INTO SOMEWHERE PRODUCTIVE AND BEAUTIFUL TO BE ENJOYED BY THE COMMUNITY.

RESPONSIBLE SEEDBOMBING

When you're launching seedbombs, there are lots of things to think about. Seedbombing responsibly includes identifying that the site you've chosen is right for the job, as well as making sure you get your aim straight – metaphorically as well as physically!

THE DOS

Identify the site

✻ Is the site protected as a conservation area? Check this out by asking the council or researching on the internet or in your local library.

✻ Is the area privately owned? Make sure the land isn't used for agricultural purposes; you don't want to interfere with food crops.

✻ Is the site abandoned and will it benefit from being beautified? Ensure the site is not due for imminent construction and that the plants will thrive there.

✻ Choose your plants wisely – non-invasive, right plant, right place. Some plants are persistent, particularly invasive, and may suffocate and/or prevent the growth of other plants (especially precious native wild flowers). We don't want another bindweed or Himalayan balsam disaster on our hands!

✻ Use native plants. Using native plants maintains the natural balance of our wildlife, flora and fauna.

✻ Swot up on your nature knowledge. For hints and tips, see **NATURE KNOWLEDGE** (page 38).

✻ Be wildlife aware. Encourage wildlife by using plants to attract bees, birds and butterflies.

GET YOUR AIM RIGHT

What is your aim? It's important to identify why you are choosing this method of gardening – it should essentially be for positive reasons. The ETHOS of seedbombing is that it's a fun and gentle form of dispersing seeds and beautifying desolate surroundings or domestic gardens.

✔ **DO** foster orphaned land and fill urban voids with flowers.

✔ **DO** grow food crops.

✔ **DO** attract wildlife.

✔ **DO** create a scenic route for both short local journeys and long motorway journeys.

✔ **DO** unite a community through plants.

✔ **DO** gain a sense of well-being.

LEFT BE SURE TO RESEARCH AREAS CAREFULLY BEFORE LAUNCHING YOUR SEEDBOMBS.

Simply drop or place seedbombs if your aiming isn't up to scratch!

ABOVE CHOOSE PLANTS APPROPRIATE FOR THE AREA, AND WHICH WON'T PREVENT OTHER PLANT GROWTH.

THE DON'TS

Identify the site

✱ Don't throw seedbombs at people, or windows. Ensure that nothing or nobody will be damaged or harmed by your flying seedbombs.

✱ Don't throw them into your neighbours' gardens without their consent. Save your seeds for something more positive than dealing with neighbourly disputes!

✱ Don't throw them on land with inadequate growing conditions. If there is insufficient light and no obvious soil for the plants to anchor themselves into, they will eventually perish.

✱ Don't put yourself or anyone else in danger when launching your seedbombs. Although it is suggested that roadsides and railways are perfect for launching your seedbombs, don't go walking across busy roads and rail tracks – it's not worth putting yourself and others at risk of an accident.

GET YOUR AIM WRONG

Your aim is wrong if you are launching seedbombs to be antagonistic or are using them as a form of vandalism or coercion. It is also wrong if you have not given consideration to the plant and its new environment.

✗ **DON'T** use seedbombs as a form of aggression or vandalism.

✗ **DON'T** launch seedbombs that are not site-specific.

✗ **DON'T** be reckless as it will have a negative impact on the community and the environment.

NATURE KNOWLEDGE

It is important to research and swot up on your nature knowledge before you plan your seedbomb launch as some areas are protected for their beauty and natural diversity, and introducing new plants – even with the best of intentions – might upset the delicate balance.

Development, pollution, climate change and unsustainable land management threaten the preservation of our wildlife and geological features and it is crucial to preserve our existing natural heritage for future generations.

UK SPECIAL SITES

(SSSI) SITES OF SPECIAL SCIENTIFIC INTEREST

A Site of Special Scientific Interest (SSSI) is a national designation giving protection to sites under the Wildlife and Countryside Act 1981 considered important for their flora, fauna and geological features. There are more than 4,000 SSSIs in England (www.naturalengland.org.uk) and they are the country's very best wildlife and geological sites.

(LNR) LOCAL NATURE RESERVE

Local Nature Reserves (LNRs) are for both people and wildlife. They offer people special opportunities to study or learn about nature or simply to enjoy it. All district and county councils have powers to acquire, declare and manage LNRs. To qualify for LNR status, a site must be considered of importance for wildlife, geology, education or public enjoyment.

(SNCI) SITE OF NATURE CONSERVATION INTEREST

A Site of Nature Conservation Interest (SNCI) is a designation used in many parts of the United Kingdom to protect areas of importance for wildlife at a county scale. In other parts of the country the same designation is known by various other names, including Site of Importance for Nature Conservation (SINC), County Wildlife Site and Site of Metropolitan Importance for Nature Conservation. Overall, the designation is referred to as a 'non-statutory wildlife site', or a 'Local Site'. The designated sites are then protected by local authorities from most development.

PLANT DIRECTORY

A BIT OF SOIL SCIENCE

Soil is a combination of organic matter and rock in its final state of decomposition after millions of years of weathering. Soil is the primary carer for plants and its ability to provide effectively depends on the make-up of the soil. Additives can improve soil and growing conditions and help plants thrive.

When you come to read the plant directory you will see that for each entry there is a 'soil requirement'. This tells you that some plants are not fussy – that is, they'll grow happily in most moist, free-draining soils – while some thrive in a certain type of soil but fail to grow at all in another.

STRUCTURE AND TEXTURE

Have you ever heard other gardeners talking about having a 'sandy loam' or maybe a 'silty clay loam' and wondered, 'What on earth are they are talking about?' After all, soil is soil, isn't it?

Not so! Soil varies widely from area to area and there are a number of factors that affect how well a plant grows in it.

The pore spaces between the particles in the soil determine how well the soil absorbs water, air and nutrients and how long the soil retains them and leaches them out.

The ideal soil is 'loamy'; it is easy to work with and retains nutrients and water while allowing it to drain.

SANDY Larger pore spaces; leaches rapidly and dries out quickly.

CLAY Smaller spaces; holds on to water and is heavy.

LOAM A mixture of sand, clay and decaying organic materials.

SILT Rock particles that are coarser than clay and finer than sand.

SOIL PH

Soil pH (potential of Hydrogen) is a measure on a scale of 0 to 14 of the soil's acidity or alkalinity.

pH is one of the most significant properties affecting nutrient availability for plants.

Some plants prefer an acidic soil. These plants are called 'calcifuge' (lime-hating plants).

Some plants prefer alkaline soils. These plants are called 'calcecoles' (lime-loving plants).

- 7 is neutral
- Alkaline is higher than pH 7
- Acidic is lower than pH 7
- 6.0 to 6.5 is the desirable range as it heightens microbial activity, allowing more nutrients to be released to the plants.
- Macronutrients are less obtainable in low pH soils.
- Micronutrients are less obtainable in high pH soils.

A BIT OF BOTANY

Some things going on within the plant are just too magical to understand. Here are some of the primary activities:

PHOTOSYNTHESIS

Plants are autotrophs – they can sustain life by making their own food.

Photosynthesis occurs primarily in the leaves; the chlorophyll in the chloroplasts captures radiant energy and generates carbohydrates and oxygen from carbon dioxide (CO_2) and water (H_2O).

CARBON DIOXIDE CO_2 is colourless, odourless and tasteless and is the main gas we exhale. Large amounts of CO_2 inhaled can be potentially lethal to humans and animals, which is why plants are so important – they use up the CO_2 in the atmosphere. When a plant dies and decomposes or is burned, CO_2 is released and returned to the atmosphere.

OXYGEN O_2 During photosynthesis, plants combine CO_2 and H_2O molecules to make carbohydrates. During this process some oxygen atoms are released.

WATER H_2O acts as a solvent for minerals, allowing their absorption and transportation within the plant. Water-filled cells keep the plant cells turgid. Plants deprived of water will wilt.

TRANSPIRATION

Water is evaporated through the leaves, a process called transpiration; it cools down the plant and opens the stomata (leaf pores used for gas exchange), which help with the intake of CO_2. Around 95% of the water taken up by the roots is transpired.

OSMOSIS

Plants take up soil water through the roots by 'osmosis', the movement of water molecules through a semi-permeable membrane in the cell walls.

Water is an electron source for the process of photosynthesis.

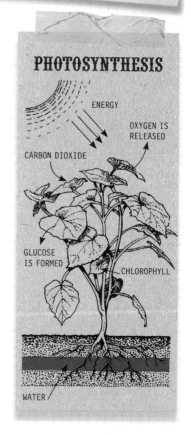

PHOTOSYNTHESIS

ENERGY

OXYGEN IS RELEASED

CARBON DIOXIDE

GLUCOSE IS FORMED

CHLOROPHYLL

WATER

NATURE KNOWLEDGE

By following a few simple maintenance procedures, you will give your plants the best chance of remaining healthy and vigorous and help to avoid attack from pests and diseases. You should act immediately upon any signs of pest damage or disease and take professional advice if necessary.

DEADHEADING

A flower's primary goal is to set seed. If you constantly cut off the dead heads the plant goes into overdrive, sending out more flowers in an effort to reproduce.

CUTTING BACK

Cutting back dead, damaged or diseased parts of the plant encourages growth and helps maintain vigour. Do this in spring or autumn and either compost the prunings or, if diseased, destroy them on a fire (use the ashes to condition the soil with potassium).

DIVISION

To divide an overgrown plant, dig it up, remove as much of the soil from the roots as you can, and cut the plant in half with a sharp knife. Replant the parent plant in the original hole and the new plant in a new position. Water thoroughly before and after planting until established.

ABOVE SUFFICIENT SPACING OF PLANTS WILL REDUCE COMPETITION FOR NUTRIENTS AND SOIL WATER, ALLOWING THEM TO THRIVE.

ROOT CUTTINGS

Take a healthy stout root, cut it into 5cm sections and pot it in well-drained compost with the tip of the cutting just shy of the top of the soil. Do not overwater your cuttings. Shoots should emerge in early spring.

AIR CIRCULATION

Plants need good air circulation to help release waste gases and reduce the chances of disease and attack. Space plants according to size, which aids circulation and prevents competing for nutrients and soil water.

Composting will also transform your kitchen or garden waste, cardbo— and paper into nutrient—r food for your garden.

FOOD FOR PLANTS

There are 13 mineral nutrient elements that have a critical function in plant growth. There are two categories – micronutrients (needed in smaller quantities) and macronutrients (needed in higher doses and administered regularly by gardeners). Most of these nutrients are found in the soil but some need to be supplemented.

NUTRI-RICH

MICRONUTRIENTS
IRON Essential for the formation of chlorophyll.
COPPER Aids reproductive growth and root metabolism.
MANGANESE/ZINC Enzyme activation.
BORON Aids sugar, carbohydrate, seed and fruit production.
MOLYBDENUM Involved in the fixation of nitrogen.
CHLORIDE Aids photosynthesis and plant metabolism.

MACRONUTRIENTS
NITROGEN (N)
PHOSPHORUS (P)
POTASSIUM (K)
CALCIUM A strengthening component of cell wall structure.
MAGNESIUM Element of chlorophyll and vital for photosynthesis.
SULPHUR Vital for plant vigour and resistance, root growth, seed production, vitamin development and chlorophyll formation.

THE PRIMARY MACRONUTRIENTS

(N) NITROGEN The plant utilizes nitrogen to produce leafy growth and for stem formation. Nitrogen requirements differ from plant to plant but as a rule the more leaf a plant produces (e.g. cabbages/spinach), the higher its nitrogen need.

Symptoms of deficiency are weak stems, stunted growth and yellowed or discoloured leaves.

Peas make good companions for nitrogen-loving plants because they fix nitrogen in the soil.

(P) PHOSPHORUS Phosphorus plays a fundamental role in seed germination and root development, especially during its infancy. It is needed throughout a plant's life for healthy root growth, fruit and seed crops.

Root vegetables such as carrots and parsnips benefit from generous doses of phosphorus.

Symptoms of deficiency are low fruit yields, stunted growth and a purple tinge to the foliage.

(K) POTASSIUM Potassium is used in the process of building sugars and starches and is needed for vegetables and fruits. It encourages flower and fruit production and is a vital nutrient for maintaining health, vigour and resistance to disease.

Potassium is found in wood ash and is perfect dug into the soil around your fruit trees and root vegetables.

Symptoms of deficiency are scorched leaves, reduced fruit yields and low resistance to disease.

AND NOW FOR THE PLANTS

This directory is an introduction to 41 fantastic plants chosen to attract wildlife and for their scent, colour, medicinal and culinary properties, and suitability for seedbombing in the wild and at home. It is by no means an exhaustive list. We have picked a wide variety of plants, from the robust teasel to the delicate poppy.

LEAF OUT OF THIS BOOK

Leaves are elegantly crafted into many different shapes and sizes with the primary goal of harvesting light. They must be sheet-like, thin and translucent (to allow light to reach the innermost cells). They must have stalks, which may develop in an opposite or alternate pattern on the stem and elevate the leaves to positions where they can track the movement of the sun throughout the day.

IN BRIEF SYMBOLS

Each entry contains a fact-file for quick reference. Here is a guide to the symbols:

✿	FAMILY
⊕	NATIVE TO
♠	HEIGHT AND SPREAD
⊁	HABITAT
✪	THRIVES IN
⚒	SOIL REQUIREMENTS
⚘	LIFESPAN
❋	FLOWERING TIME
⚘	FORM
◆	LEAF FORM
🐝	POLLINATED BY
✏	NOTE*
⚘	CONSERVATION STATUS

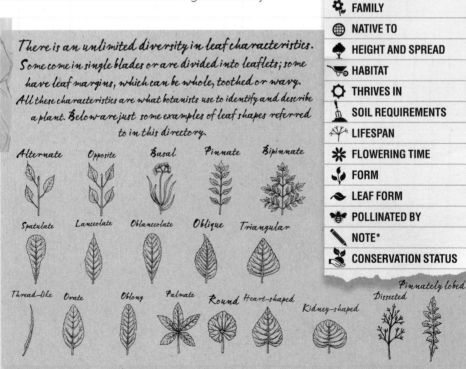

There is an unlimited diversity in leaf characteristics. Some come in single blades or are divided into leaflets; some have leaf margins, which can be whole, toothed or wavy. All these characteristics are what botanists use to identify and describe a plant. Below are just some examples of leaf shapes referred to in this directory.

Alternate Opposite Basal Pinnate Bipinnate

Spatulate Lanceolate Oblanceolate Oblique Triangular

Thread-like Ovate Oblong Palmate Round Heart-shaped Kidney-shaped Dissected Pinnately lobed

WILD CHAMOMILE
Matricaria chamomilla

Chamomile was cultivated as early as the Neolithic period and has been used for centuries as a 'cure-all' medicinal plant. A great companion plant as its strong, aromatic flowers attract beneficial insects that feed on pest predators, such as aphids.

IDENTIFICATION

STEMS Branched, upright, smooth stem.

LEAVES The long and narrow alternate leaves can be harvested fresh from the plant for medicinal uses.

FLOWERS Yellow and comb-like centres (often likened to a pineapple) and surrounded by 10–20 white petals. Harvest the flowers for medicinal uses when open and fresh or dry for later use. It takes 20–35 days from flower to seed; the petals bow down and wither as the fruit is forming.

SEEDS 1mm elongated, light brown and ridged.

LAUNCH SEEDBOMBS April to May and August to September.

GERMINATION TIME 1–2 weeks.

HARVESTING SEEDS Seeds ripen July to September.

PLANT CARE Don't cut back the foliage before flowering as the flower production will reduce dramatically. • Aphids adore chamomile – to remove them, wash off with a strong jet of water.

CULINARY AND MEDICINAL USES
Medical Chamomile has calming and soothing properties and is used to treat nervousness, anxiety, hysteria, headaches, stomach pains, indigestion, colds and flu. Also used as a poultice for swellings, sprains and bruises. • Steep for 15 minutes then drink for a gentle sleep aid.

IN BRIEF

✿	Asteraceae/Compositae
⊕	Southern Europe
♣	60 x 45cm
🌾	Roadsides, railways, waste ground, fields, arable land
☼	Full sun/partial shade
⚒	Most soil types; tolerates poor soils
⚘	Annual
❋	May to August
⚘	Upright
⬳	UPPER: Bipinnate; LOWER: Tripinnate
🐝	Insects
✎	Few allergic reactions to chamomile have been reported
🌱	No known conservation issues

COMMON POPPY
Papaver rhoeas

The poppy is one of the most commonly recognized wild flowers. During World War I, common poppies bloomed in waste grounds and served as a vivid reminder of the battles that took place; they are now an iconic flower to military veterans as the symbol of remembrance.

IN BRIEF

✿	Papaveraceae
🌐	Europe
♠	60 x 20cm
🛒	Roadsides, railways, waste ground, fields, cultivated beds
☼	Full sun
⚒	Grows on most moist free-draining soils (can suffer on heavy clay)
⚘	Annual
✳	June to August
🌱	Wiry, clump-forming
➤	Pinnately divided
🐝	Beetles, bees, flies
✎	The latex in the seed pods is a mild narcotic and slightly sedative. The seeds are NOT toxic
🌿	No known conservation issues

IDENTIFICATION

STEMS The stems grow from a big taproot and are upright, wiry and branching; they are green with a purplish tinge and have tough hairs.

LEAVES Alternate divided, serrated pairs of narrow, toothed leaves.

FLOWERS Hermaphrodite, four to six vivid red, wrinkled petals with a black blotch at the base. Yellow and brown anthers radiate from the central ovary; the ovary ripens and sheds seeds 3–4 weeks after flowering.

SEEDS The seed pod is small, green/blue and shaped like a pepperpot – as the seed pod dries and browns, the holes open around the top edge, and as the wind blows the seeds are released from the holes. They are tiny round black dots, which, close up, are actually slate-blue and kidney-shaped. The average number of seeds per seed pod is 1,360.
• The number of seeds per plant ranges from 10,000 to 60,000. The seeds can remain viable in the soil for up to 8 years.

LAUNCH SEEDBOMBS March to April and September to October.

GERMINATION TIME 1–6 weeks.

HARVESTING SEEDS Seeds ripen for harvesting between August and September.

PLANT CARE Poppy self-seeds readily and may need deadheading to prevent prolific spread (use this procedure as part of your seed-harvesting routine). • As long as they have sunshine they are happy.

The poppy has a long flowering season, from June to October. It makes an attractive cut flower, though the petals drop after a few days.

• Poppies thrive best in disturbed land; give your seeds a helping hand at germinating by annually digging over the spot where they have fallen.

PESTS AND DISEASES

Pests Can suffer attacks from greedy aphids.
Diseases Can be susceptible to downy mildew.

CULINARY AND MEDICINAL USES

Edible Ancient Egyptians pressed the seed to obtain cooking oil, which was used with honey to create a sweet confection.
• The seeds have a nutty flavour and are sometimes used for baking breads, cakes and biscuits.
Medical Common poppy has been used as a mild pain reliever and a sedative and to treat irritable coughs, sleeplessness and digestive problems. • Common poppy is not addictive like its relative the opium poppy, but the seed pods do contain alkaloids and should only be used under the supervision of a qualified herbalist.

OTHER USES A red dye is obtained from the flowers, which can be used to dye foods and medicine syrups.
• An infusion can be made from the petals that, when applied to the skin, may reduce wrinkles.
• Dried petals can be added to potpourri.
• Specialist flower essenses can be prescribed to help balance dysfunctional emotions.

The poppy is a valuable foodplant for wildlife.

RIGHT SEEN AS AN AGRICULTURAL WEED, THE POPPY ENJOYS GROWING ON REGULARLY DISTURBED LAND.

CORN COCKLE
Agrostemma githago

Corn cockle is easy to grow in any position and ideal for naturalizing in a wild-flower meadow. It is a quick grower and produces slender stems carrying mauve flowers, which attract birds, bees and butterflies (the flowers close at night). The whole plant is delicately covered with fine silver hairs.

IN BRIEF

🌸	*Caryophyllaceae*
🌐	Europe
♠	120 x 30cm
🛒	Roadsides, railways, waste ground, fields
☼	Full sun/partial shade
🌱	Tolerates most moist soils
⁂	Hardy annual
❋	June to August
🌱	Upright, slender
➤	Lanceolate
🐝	Long-tongued insects like butterflies, moths and bees
✎	All parts of the plant are poisonous if ingested
🌿	Intensive mechanized farming has put the plant at risk and it is now virtually extinct in the wild

IDENTIFICATION

STEMS Fuzzy, branched, slender stems.

LEAVES The stem is lined with opposite hairy, narrow lanceolate, grey-green leaves.

FLOWERS Delicate five-petalled pink scentless flowers. Each petal bears two or three irregular black lines. There are five narrowly pointed green sepals; these exceed the petals and are joined at the bottom to form a downy furrowed tube, which develops into a seed capsule.

SEEDS Many seeds are produced in the capsule – quite possibly the coolest seed I've ever seen!

LAUNCH SEEDBOMBS May and September.

GERMINATION TIME 1 week.

HARVESTING SEEDS Collecting the seed early in the season avoids seed-borne disease.

PLANT CARE This upright fast-growing plant needs support with stakes to avoid wind damage. • Deadhead to prolong flowering and encourage more flower buds. • Needs regular irrigation but suffers if waterlogged. • Fertilize with NPK when the plant is juvenile. • Generally pest- and disease-free.

CULINARY AND MEDICINAL USES
Edible Not recommended for consumption.
Medical Not recommended for medicinal use.

RED CAMPION
Silene dioica

Red campion is a relatively short-lived perennial.
The male and female flowers are borne on separate
plants. Often associated with woodlands and country
roadside verges but can be grown in cultivated beds.
It has a long flowering period right through to autumn.

IN BRIEF

🌸	*Caryophyllaceae*
🌐	Europe
🌳	90 x 30cm
🛒	Roadsides, railways, waste ground, fields, maritime, cliffs and seabird crevices
☀	Full sun/partial shade
🏺	Drought tolerant; grows on most soils but thrives on free-draining, moist, calcareous soils
🌱	Herbaceous perennial
❄	May to October
🌿	Clump-forming
🍃	Ovate
🐝	Bees, flies
✏	No known hazards
🌿	No known conservation issues

IDENTIFICATION

STEMS Wiry, strong, upright branched, round downy stems.

LEAVES Deep green in opposite paired, slightly sticky leaves.

FLOWERS Downy, somewhat sticky branching stems hold dioecious unscented rose pink flowers, which consist of five deeply notched petals joined at the base to form an urn shape and surrounded by a purple/brown calyx.

SEEDS The seed pod is an ovoid capsule containing numerous seeds. When dry, the seed pod opens at the apex as several teeth curve back to release the seeds. These are kidney-shaped, 3mm long, and the colour varies from red/brown to black. Microscopically, they have a wonderful bobbly texture.

LAUNCH SEEDBOMBS Any time of year.

GERMINATION TIME 1–2 weeks.

HARVESTING SEEDS Seed ripens July onwards. The female plant produces thousands of seeds per season, which can remain viable for many years.

PLANT CARE Red campion is a low-maintenance plant and may only need autumn dividing if it has become too big.
• Relatively pest- and disease-free. • Suffers with prolonged periods of waterlogging.

CULINARY AND MEDICINAL USES
Edible No known culinary uses.
Medical Not regarded as a useful medicinal herb.

BETONY
Stachys officinalis

Supposedly taken from the Celtic word *bewton*, meaning 'good for the head', Betony is said to have been one of the most important medicinal plants for Anglo Saxons of early medieval Britain. Betony is a slow-growing, long-lived hardy perennial found in dry grassland, meadows, open woods and slopes.

IN BRIEF

🌼	*Lamiaceae/Labiatae*
🌐	Europe, western Asia and North Africa
🌳	70 x 45cm
🛒	Roadsides, railways, waste ground, woodlands, grassy cliff-tops, grassy banks
☼	Full sun/dappled shade
🛠	Tolerates poor soils and thrives on damp soils
🌳	Hardy perennial
❉	July to September
🌱	Hardy perennial
🍃	UPPER: Lanceolate; LOWER: Oblong/ Heart-shaped
🐝	Bees
✎	Not recommended during pregnancy
🌿	Rare in Ireland and Northern Scotland; it is classified as Endangered and is protected by the 1999 Flora Protection Order

IDENTIFICATION

STEMS Like all plants in the *Lamiaceae* family, *Stachys officinalis* has distinctive square purplish stems with a branching leaf axis.

LEAVES Pairs of alternate, stalkless, narrow, ovate, fluffy leaves are borne sparsely towards the top of the stem. • Most of the foliage is formed at the base of the stem and the leaves are larger, heart-shaped and rough to the touch with fine, short, fluffy hairs and toothed margins. The entire surface of the leaf has glands containing bitter, aromatic oil.

FLOWERS Short flower spikes grow on the top of the stem; further groups of flowers grow at intervals up the stem, a characteristic of betony. • The rich purplish/pink hermaphrodite flowers are tubular and two-lipped; the upper lip forms a hood and the lower lip has two short side lobes and a large wide middle lobe.

SEEDS The seeds are 0.3mm long and rusty/brown coloured, with an angular shape.

LAUNCH SEEDBOMBS March to September or October to February because the seeds need a period of chill to break their dormancy and speed up germination.

GERMINATION TIME Irregular – often takes between 30 and 90 days.

HARVESTING SEEDS Seeds ripen from July to October. • Cut the stems several inches above the soil line and strip the leaves off.

PLANT CARE Betony is frost-hardy and drought tolerant. • Ensure good air circulation and drainage and occasionally weed around the plant to allow it to spread.

• Divide mature clumps in the spring or autumn to increase health and vigour and make more plants.

PESTS AND DISEASES

Pests Young growth can be attacked by slugs.
Diseases Seldom attacked by diseases but can suffer from root rot caused by a fungus present in boggy conditions.

ABOVE BETONY IS COMMON IN ENGLAND AND WALES, BUT RARE IN IRELAND AND NORTHERN SCOTLAND.

CULINARY AND MEDICINAL USES

Edible The leaves and flowering tops make a good caffeine-free substitute for tea. Brew in water to taste. Betony tea has all the good qualities that normal tea has and bypasses all the bad ones. Historically taken before a drinking session to decrease the chances of a hangover!
Medical Modern herbalists prescribe Betony to treat conditions such as high blood pressure, anxiety, heartburn, migraine and stress headaches, and as an antiperspirant.

OTHER USES An ointment can be made to soothe skin complaints.

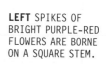

LEFT SPIKES OF BRIGHT PURPLE-RED FLOWERS ARE BORNE ON A SQUARE STEM.

CORNFLOWER
Centaurea cyanus

Cornflower was once a common weed of cornfields and arable land but due to modern intensive farming practices and the use of herbicides its presence in the wild has been drastically cut down. With their attractive flower heads in rich shades of blue, they are much sought-after for gardens and flower arrangements.

IN BRIEF

	Compositae/Asteraceae
	Europe
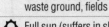	90 x 30cm
	Roadsides, railways, waste ground, fields
	Full sun (suffers in shade)
	Thrives on most soils; tolerates poor soils (suffers in drought)
	Hardy annual
	June to August
	Slender
	Lanceolate
	Long-tongued insects like butterflies, moths and bees
	Poisonous to cats if ingested
	In 1990 Cornflower sightings dramatically declined in the UK. Various charities are working towards preventing extinction

IDENTIFICATION

STEMS The flowers appear at the top of downy grey/green slender hollow stems on plump, hard involucres with bracts resembling overlapping tiles with black jagged edges.

LEAVES Lanceolate alternate, stalkless, downy grey-green leaves.

FLOWERS Cornflower has a compound head of ruffled vivid blue tubular flowers; on the outer edge sit the large blue flowers and in the middle of the flower head rest the smaller, brilliant purple flowers.

SEEDS The seed head opens as it dries to reveal many seeds surrounded by soft silver fluff, which aids wind dispersal. The seeds are 5mm long and have a pearly white, shiny colour with a tuft of light brown bristles.

LAUNCH SEEDBOMBS March to May but autumn launching produces larger, earlier-flowering plants.

GERMINATION TIME 1–2 weeks.

HARVESTING SEEDS Harvest cornflower seeds throughout summer as the flower dies; it can be incorporated as part of the deadheading routine.

PLANT CARE The wiry plants may need some support to prevent them from collapsing or from disorderly growth. • Deadheading helps to extend the flowering season. • Cut the faded flower heads down to the ground in autumn, and compost the spent brown stems. • Can be invasive so cut seed heads off before self-seeding and occasionally divide and thin out the plant.

PESTS AND DISEASES

Pests Cornflowers are relatively pest-free apart from one bothersome pest – the aphid!

Diseases Cornflowers can suffer from rust and from powdery mildew.

CULINARY AND MEDICINAL USES

Edible The blue dye obtained from the petals is edible and used for colouring sweeties. The young shoots and flowers can be used in salads and as a garnish.

Medical Cornflower has a long history of use as a medicinal herb, though it is rarely used in modern medicine. A tonic can be made and used as a mild purgative for the treatment of constipation and to improve the digestion, or as a mouthwash for bleeding gums and mouth ulcers.

External Used as an anti-inflammatory and astringent herb for skin cleansing, the infused blossoms can be made into an eyewash for such ailments as conjunctivitis and to relieve puffy eyes. The leaves can be used in a facial steam for dry skin.

OTHER USES A blue dye is obtained from the petals and when mixed with alum-water is used to dye linen and wool. • The dye is also added to cosmetic products like shampoo and hair dye – the famous blue rinse!

ABOVE CORNFLOWERS GROW IN SUNNY ARABLE LAND, ESPECIALLY CORNFIELDS – HENCE THE NAME.

FOOD FOR ALL

They attract butterflies and bees, which feed from the nectar, and birds for the seed heads.

These annuals will flower profusely throughout the summer months.

LEFT CORNFLOWER IS A HARDY PLANT WITH A SLENDER HABIT, BRANCHED STEMS AND GREY/ GREEN LEAVES.

LESSER KNAPWEED

Centaurea nigra

Considered a weed in Australia and the USA, this wild flower is common in Europe. It grows in large colonies and is a vital foodplant for birds, bees, butterflies and moths as it is nectar-rich and provides autumn and winter seeds for many birds.

IN BRIEF

🌸	*Asteraceae*
🌐	Europe and introduced into US and Australia
🌳	65 x 30cm
🛒	Roadsides, railways, waste ground, cliffs, grassland
☀	Sun and semi-shade; tolerant of maritime conditions
🌱	Drought-resistant; thrives in most soil types; requires well-drained soil; tolerates poor soil
🌿	Hardy herbaceous perennial
❄	June to September
🌿	Clumps of upright stems
🍃	UPPER: Lanceolate; LOWER: Lobed
🐝	Long-tongued insects like butterflies, moths and bees
✏	No known hazards
🌱	Rare in some parts of Scotland

IDENTIFICATION

STEMS Tough, wiry, prostrate, woody, roughly hairy and grooved stems.

LEAVES The leaf shape throughout the plant varies; the upper leaves are dull green, stalkless, downy and lanceolate. The lower leaves are stalked and deeply lobed, some with coarse teeth, and are often confused with the thistle.

FLOWERS Lesser knapweed flowers are hermaphrodite and are purple, thistle-like and come in two forms, rayed and un-rayed. The flower sits on a hard head of overlapping bracts.

SEEDS As the hard seed head dries, it opens to release around 60 seeds to be dispersed by the wind; they can travel up to many miles. The seeds are 5mm long, pearly white with lengthwise stripes, little notches at the base and a tuft of short light brown bristles at the top.

LAUNCH SEEDBOMBS Early spring through to autumn, but this is such a brilliant plant and if conditions are suitable it will germinate and grow at any time of the year.

GERMINATION TIME 1–2 weeks.

HARVESTING SEEDS Seeds ripen for harvesting from August to October. Collect when plump and hard and all the petals are absent.

PLANT CARE Leaving the seed heads uncut after flowering provides food for garden birds, but there will also be a risk of unwanted spread of the plant through self-seeding. • Lesser knapweed is an easy plant to grow and once established can tolerate considerable amounts of neglect. • It can be invasive as it will self-seed freely, so cut seed heads off early. • Divide the plant in autumn once every three years to maintain its health and vigour. • The plant will suffer in damp and in acid soils.

PESTS AND DISEASES

Pests Knapweed is a tough cookie and relatively pest-free, the only threat being birds eating the seeds, but there are more than enough seeds to go around!

Diseases Seldom attacked by any diseases.

CULINARY AND MEDICINAL USES

Edible The flowers petals can be eaten raw, sprinkled over salads or used as a garnish.

Medical Historically used as a diuretic and tonic, and to treat cancer; made into an ointment, it was used externally to heal wounds and skin ailments such as cuts and bruises. It was also used to soothe sore throats and bleeding gums.

ABOVE THIS HARDY HERBACEOUS PERENNIAL IS A POPULAR FOODPLANT FOR WILDLIFE.

Very self-sufficient, they require little care to flourish in the wild.

LEFT ALTHOUGH CONSIDERED A WEED, HENCE THE NAME, LESSER KNAPWEED IS ACTUALLY A TREMENDOUSLY ATTRACTIVE AND VERY LOW-MAINTENANCE PLANT.

TO DYE FOR

A yellow dye is obtained from the flowers. They are also often used for attractive fresh and dry floral arrangements.

FIELD SCABIOUS
Knautia arvensis

A tap-rooted herbaceous perennial that soars above its neighbouring plants. The lower leaves are thistle-like, while the upper are lance-shaped. It receives regular visits from bees, butterflies, moths, hoverflies and birds.

IN BRIEF

	Dipsacaceae
	Europe, Asia
	10 x 30cm
	Roadsides, railways, waste ground, fields
	Full sun (suffers in deep shade)
	Drought tolerant; grows on most soils but thrives on free-draining calcareous soils
	Hardy perennial herb
	July to October
	Upright, clump-forming
	UPPER: Pinnately lobed; LOWER: Lanceolate
	Long-tongued insects like butterflies, moths and bees
	No known hazards
	No known conservation issues

IDENTIFICATION

STEMS Round stems, bare of leaves, branched and coarsely covered in whitish hairs.

LEAVES Dull green leaves form a basal rosette.

FLOWERS Delicate rounded flat flower heads are usually in shades of purple/mauve and comprise around 50 tiny, densely packed, four-lobed flowers with protruding pink anthers (hence its likeness to a pin cushion).

SEEDS The seeds are achenes, and are light brown, cylindrical, slightly hairy and 5mm long. They fall to the ground when they are ripe.

LAUNCH SEEDBOMBS March and September.

GERMINATION TIME Erratic, up to 30 days.

HARVESTING SEEDS August to October.

PLANT CARE Field scabious is an invasive species and may need controlling if grown in a cultivated bed – thin out by digging up in the autumn and destroying. • The plant overwinters as a rosette of dark green leaves. • Cut off spent flower heads and compost them.

CULINARY AND MEDICINAL USES
Edible No known culinary uses.
Medical Historically has been used as a medicinal herb to treat skin disorders, cuts, burns, bruises and boils, and also for blood purification.

WILD TEASEL
Dipsacus fullonum

A biennial plant, teasel forms a prickly rosette of leaves in the first year of growth and the flowering stem emerges in the following growing season. Common names include prickly beehives and church brooms.

IN BRIEF

🌸	Dipsacaceae
🌐	Europe, North Africa, Asia
🌳	200 x 80cm
🛒	Roadsides, railways, waste ground, fields, riverbanks
☀	Full sun/partial shade
🏺	Most free-draining soils
🌱	Biennial
❋	July to August
🌱	Upright, clump-forming
🍃	BASAL: Ovate; UPPER: Lanceolate
🐝	Bees
✏	No known hazards
🌿	No known conservation issues

IDENTIFICATION

STEMS Stout upright stems are produced during the second year of growth. Downward prickles grow in ridges along the stem.

LEAVES The first season of growth produces a basal rosette of pale green prickly leaves. The leaves on the flowering stem form a kind of bowl-shape, which collects water.

FLOWERS The spiny cone-like flower heads are 3–10cm long and are covered in tiny hermaphrodite purplish tubular flowers, 10–15cm long, occurring in a ring around the head. The base of the flower head has several leaf-like bracts curving upward.

SEEDS There are approximately 2,000 flowers per flower head; each one produces a 4–5mm long, sandy-brown furrowed seed.

LAUNCH SEEDBOMBS February to May and August to October.

GERMINATION TIME 1–4 weeks.

HARVESTING SEEDS August to October.

PLANT CARE Because they seed freely they can grow in unwanted places; to prevent this, pull out unwanted seedlings.
• As the plants die down they leave a large area of bare ground ready for the seeds to germinate into new plants. • Will withstand harsh weather conditions and do not require staking.

CULINARY AND MEDICINAL USES

Edible No known edible uses.

Medical Used to treat fractures as it contains nutrients that strengthen bones, cartilage and sinews. Also used to treat stomach complaints, and to promote blood circulation and energy.

GREATER HAWKBIT
Leontodon autumnalis

Considered a weed, this wild flower is in the same family as dandelion. It has a long flowering season, provides late-season colour and nectar for a variety of insects, and is a foodplant for birds. It's extremely hardy, and able to grow on sites with regular disturbance such as cut and grazed fields.

IN BRIEF

🌸	Asteraceae/Compositae
🌐	Europe
🌱	30 x 30cm
🛒	Roadsides, railways, waste ground, fields, shores
☀	Full sun
🌱	Drought tolerant; grows on most soils
🌲	Hardy perennial herb
✳	June to October
🌿	Sprawling/low
🍃	Lanceolate
🐝	Long-tongued insects like butterflies, moths and bees
✎	Prolonged ingestion of the herb could cause side effects
🤲	No known conservation issues

IDENTIFICATION

STEMS The stems are thin, tough, branched and round.

LEAVES A low-growing basal rosette of long, deeply cut narrow leaves.

FLOWERS Yellow 'composite' flat flower with square-tipped petals with a serrated edge and a reddish-tinged underside. A slightly hairy involucre rests directly below the petals; it is pale green with yellow reddish tinges.

SEEDS The seeds are cylindrical, smooth, light brown, with a feather-like pappus of branching hairs.

LAUNCH SEEDBOMBS Any time of the year.

GERMINATION TIME 1–3 weeks.

HARVESTING SEEDS Once pollinated, the petals close and the fruit forms. Cut off the seed head before it opens to release the seeds.

PLANT CARE A hardy plant that needs little attention. • Greater hawkbit relies upon seed for regeneration so any unwanted repopulation of the plant can be managed by clipping the flowers off before they seed.

CULINARY AND MEDICINAL USES

Edible Fresh leaves can be added to salads and the roasted root made into tea and a coffee substitute.

Medical For many centuries the plant has been used as a diuretic, to detoxify and to improve bone health.

MUGWORT
Artemesia vulgaris

Regarded as a roadside weed, this rhizomous aromatic perennial was known in ancient times as one of the most powerful medicinal herbs. A valuable foodplant for Lepidoptera, which feed on the leaves and flowers.

IN BRIEF

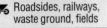
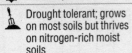

- Asteraceae/Compositae
- Europe, Asia, North Africa, North America
- 200 x 70cm
- Roadsides, railways, waste ground, fields
- Full sun/partial shade
- Drought tolerant; grows on most soils but thrives on nitrogen-rich moist soils
- Herbaceous perennial herb
- July to September
- Bushy, dense, spreading
- Pinnatipartite to Bipinnate
- Wind
- Not recommended during pregnancy or while breast-feeding. Prolonged ingestion could cause side effects
- No known conservation issues

IDENTIFICATION

STEMS The upright stem has a red-purplish tinge.

LEAVES The dark green, deeply indented, pinnate, feather-like leaves have a downy white underside.

FLOWERS Panicles of 5mm-long reddish/yellow hermaphrodite flowers.

SEEDS Each stem produces up to 9,000 glabrous seeds.

LAUNCH SEEDBOMBS February to March.

GERMINATION TIME 1–2 weeks.

HARVESTING SEEDS August to October.

PLANT CARE Mugwort is vigorous and doesn't require any special attention. • Deadhead before it sets seed. • Note: It has growth inhibitors in its roots that may reduce vigour of neighbouring plants.

CULINARY AND MEDICINAL USES

Edible All plant parts can be used fresh or dried. Its leaves have a mild sage aroma and have been used to flavour beer and meat and fish dishes. Young leaves can be boiled as a pot herb or used in salad. Pick the leaves and buds shortly before it flowers.

Medical Historically it has been used as an anti-inflammatory and antiseptic, and for digestive problems, coughs and colds, fevers and gynaecological problems.

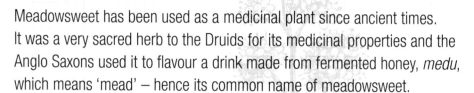

MEADOWSWEET
Filipendula ulmaria

Meadowsweet has been used as a medicinal plant since ancient times. It was a very sacred herb to the Druids for its medicinal properties and the Anglo Saxons used it to flavour a drink made from fermented honey, *medu*, which means 'mead' – hence its common name of meadowsweet.

IN BRIEF

 Rosaceae

 Europe, Asia

 120 x 45cm

 Roadside ditches, riverside, woodland, meadows, wild gardens, cultivated beds

 Full sun/partial shade

 Most moist, well-drained soils

 Hardy herbaceous perennial herb

 June to August

 Clump-forming, upright

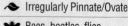 Irregularly Pinnate/Ovate

 Bees, beetles, flies

 Not to be ingested if you are allergic to aspirin, and not recommended during pregnancy or for nursing mothers or children under 12 years

No known conservation issues

IDENTIFICATION

STEMS The stems are tall, upright/arching, reddish-purple, angular and furrowed; it branches nearer the top and bears long petioled alternate leaves.

LEAVES Fern-like, deeply veined leaves comprising two to five pairs of ovate, serrated, dark green, almond-scented leaflets with downy undersides. There are three to five terminal leaflets at the end of each leaf.

FLOWERS The flowers are sweetly scented, hermaphrodite and composed of panicled cymes of creamy-white flower clusters; each flower has five petals and over 20 protruding stamens, giving the flowers a delicate fuzzy appearance.

SEEDS The seeds are 1mm long, tan-coloured (when dry) and are formed in globe-like green clusters of six to ten seeds. The seeds can be dispersed via water and can float for several weeks before bedding in the riverbank.

LAUNCH SEEDBOMBS March to September.

GERMINATION TIME Up to 4 weeks.

HARVESTING SEEDS The seeds ripen from August to September and can be sown green.

PLANT CARE Meadowsweet has a tendency to self-seed everywhere; to control this, remove the faded flowers before the seeds form. • Divide in spring or autumn. • Meadowsweet likes to be watered copiously and regularly. • The plant is deciduous and dies down in the winter; after the birds have fed from the seeds, cut back in (autumn) to prevent frost damage.

PESTS AND DISEASES

Pests Generally pest-free.

Diseases Can suffer from powdery mildew and rust fungus.

CULINARY AND MEDICINAL USES

Edible Meadowsweet flowers have been used to flavour hot, cold and alcoholic beverages and made into syrup for fruit salad. The leaves flavour soups and stews.

Medical Meadowsweet gently relieves pain and can be used as an anti-inflammatory and to treat digestive and diuretic problems, heartburn, headaches, menstrual cramps, common colds, sickness and rheumatic pain. • A decoction can be made to use externally as a wash for wounds and to soothe sore eyes.

ABOVE MEADOWSWEET IS A VALUABLE FOODPLANT FOR MOTH CATERPILLARS, INSECTS, BEES AND BIRDS.

LEFT THE FLOWER HEADS ARE RICH IN SALICYLIC ACID, A CHEMICAL THAT LED TO THE DEVELOPMENT OF ASPIRIN, WHICH WAS FORMERLY NAMED ACETYLSALICYLIC ACID.

OTHER USES A black dye can be obtained from the roots and a yellow dye from the plant tops. • The flower buds provide an essential oil, used in cosmetics and perfumery. • Gather the flowers during the flowering period and use as a cut flower or dry and add to potpourri.

WET, WET, WET

Meadowsweet is a delicate, graceful perennial herb with strongly scented creamy white flowers. It is common throughout Europe and found in the USA and Canada. It likes to grow in damp meadows, woodland and by rivers; it will even tolerate boggy conditions, though it prefers growing where moisture fluctuates as it suffers if the ground is constantly waterlogged.

COWSLIP
Primula veris

Cowslips are a well-loved English countryside wild flower. Their populations declined in the last century, but having been recorded as a protected species, and due to changes in agricultural practice, they are staging a revival. They are increasingly appearing on motorways and roadsides.

IN BRIEF

- ❀ *Primulaceae*
- 🌐 Europe
- 🌱 25 x 25cm
- 🛒 Roadsides, motorway banks, railways, waste ground, fields, meadows, maritime, woodland, lawns
- ☼ Full sun/partial shade
- 🛠 Most moist, nutrient-rich, free-draining soils
- ⚘ Hardy herbaceous perennial
- ✳ April to May
- ☘ Low-growing
- ◗ Ovate
- 🐝 Long-tongued insects like butterflies, moths and bees
- ✎ Not recommended during pregnancy
- 🌿 Absent in northerly areas including much of north-west Scotland. Should not be collected in the wild

IDENTIFICATION

Cowslips grow in clumps very close to the ground. The first emerging leaves are tight coils, which unroll to form a rosette from which emerge stocky, round, pale green, single-flowering stems.

STEMS The cowslip's stem is stout, light pale green, fuzzy/downy, upright and round.

LEAVES The leaves are oval, fresh green, crinkly, toothed and tough, covered with downy hairs and around 15cm long and 6cm wide, with paler green to white midribs and veins.

FLOWERS Hermaphrodite clusters (umbel) of 10–30 funnel-shaped, orange-based, deep yellow nodding flowers, 9–15mm long, with five heart-shaped petals that usually droop to one side of the stem and are sweetly fragrant. The flowers are followed by clustered seed pods.

SEEDS The seed pod is an ovoid capsule containing several brown/black seeds, which take a while to ripen and dry; eventually in July they open at the apex, where several teeth curve back to release the seeds.

LAUNCH SEEDBOMBS July – If collected and sown while still green, it will not go dormant and will germinate rapidly.
• September – Will be dormant already and need the cold winter months to break dormancy, which will slow germination time.

GERMINATION TIME 6 months (the following spring).

HARVESTING SEEDS Seed pods ripen from July to August and can be sown green or dried for later use.

PLANT CARE If grown in a lawn, don't mow around the plants until they have finished flowering. Cowslip will self-seed on the lawn; the seedlings can be pricked out and grown on in pots.
• Suffers if waterlogged – make sure it's planted in a free-draining site. • Divide from September after flowering. • Cut off and compost any browning or yellowed leaves.

PESTS AND DISEASES
Pests Caterpillars may be found around and underneath the leaf rosette in springtime. • Aphids may attack the soft tissue of the plant.
Diseases Mosaic virus, crown rot, downy mildew and leaf spot.

CULINARY AND MEDICINAL USES
Edible Cowslip leaves have been used to make wine and vinegar, in salads, and cooked in soups and as a spinach substitute. • The dried leaves can be made into a tea, and the flowers can be made into a jam conserve.
Medical Cowslips have a long history of medicinal use to treat conditions concerning paralysis and rheumatic pains, spasms and cramps. They have also been used as an expectorant, diuretic and anti-inflammatory and to treat chronic coughs, fevers and sleeplessness.

BEE-AUTIFUL
Rosettes of crinkly leaves emerge in late winter and are followed by deep yellow spring flowers, which provide an early nectar source for nectar-loving insects such as bees.

RIGHT DELICATE, YELLOW, CHEERFUL FLOWERS ARE BORNE ON STOUT STEMS, WHICH RISE FROM A ROSETTE OF RICH GREEN LEAVES.

Cowslips are good grown in drifts or colonies and can be naturalized in lawns.

RED CLOVER
Trifolium pratense

Red clover is a wild plant with a sprawling habit, usually found in fields and meadows and unattended wasteland. It is used as grazing fodder for livestock for its high protein content and high yields. It also has a number of medicinal benefits for ailments.

IN BRIEF

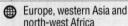

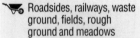

- Leguminosae
- Europe, western Asia and north-west Africa
- 10–45 x 60cm
- Roadsides, railways, waste ground, fields, rough ground and meadows
- Sun/partial shade (suffers in full shade); tolerates strong winds but suffers from maritime exposure
- Thrives on chalky soils but will grow in most soil types; requires moisture
- Hardy perennial herb
- May to September
- Sprawling clumps
- Trifoliate with a white V-shaped chevron on each leaf
- Bees
- Food for wildlife and livestock
- No threats or conservation issues

IDENTIFICATION

STEMS Its rhizomatous roots send out runners, which produce numerous sprawling, herbaceous, slightly hairy, round stems up to 60cm high.

LEAVES Clover has trifoliate leaves which usually come in threes, though it does vary on rare occasions. The 2cm-long ovate leaves are slightly pointy, hairy/toothed, and have a white chevron marking in the centre. The leaves are food for moths and butterflies.

FLOWERS Hermaphrodite, red/purple, dense balls of fragrant flowers are borne between two leaves. The flowers remain vertical when awaiting fertilization and will droop when they have been fertilized by a visiting bee.

SEEDS Its smooth-coated, kidney-shaped seeds are up to 2mm long and vary in colour such as tan, brown, red, yellow and green. The seeds are an important food source for birds. Seeds ripen from July to October.

LAUNCH SEEDBOMBS In spring or late summer when there is sufficient soil moisture for seed germination.

GERMINATION TIME 1–14 days.

HARVESTING SEEDS Harvest the seeds when the flower heads have turned black (30 days after full bloom).

PLANT CARE Clover requires little attention to grow vigorously. • Thrives on most well-drained soils and can tolerate nutritionally poor soil. • Plant in an open sunny spot. • Clover suffers in heavy shade and in maritime conditions. • Over-watering could cause the plant to rot.

PESTS AND DISEASES
Pests Slugs will destroy clover seedlings! – and red clover mites are a minor threat.
Diseases Clover rot.

CULINARY AND MEDICINAL USES
Edible The flowers have a sweet flavour and the leaves are strongly flavoured with a crisp texture; they can be added to soups, salads and as a garnish.
Medical Red clover tea helps calm an upset stomach, soothes coughs, nerves and sickness and relaxes the muscles. Add 1 tbsp of fresh or dry flowers to a cup of boiling water, brew for 10 minutes and add honey to taste (note: also suppresses appetite). • Red clover has many nutrients including Vitamin C, calcium, copper, zinc, magnesium, niacin, phosphorous, potassium, thiamine and chromium and is rich in isoflavones.

OTHER USES Red clover ointments have been used to treat skin complaints such as psoriasis, eczema and other rashes.

NOTE Avoid gathering clover from agricultural land as it may have been treated with herbicides. Grow your own!

FOLKLORE In the Middle Ages, clover was considered a charm for protection and good luck.

ABOVE CLOVER MAKES FANTASTIC LOW-MAINTENANCE GROUNDCOVER AND IS A FIELD FAVOURITE PLANT.

Red clover requires little care and will spread rapidly in the wild.

LEFT RED CLOVER'S CHARACTERISTIC THREE LEAVES AND POMPOM PINK FLOWERS MAKE IT A WIDELY RECOGNIZED PLANT.

SOIL FOOD

Like all plants in the *Leguminosae* family, clover has a symbiotic relationship with certain soil bacterias that form nodules on the roots and fix atmospheric nitrogen, which conditions the soil.

LADY'S BEDSTRAW
Galium verum

The common name was given because historically it was used as bedding for its coumarin scent, which combated sleeplessness and repelled fleas. It's related to Cleavers and is a good groundcover plant for neglected areas. Noted for attracting wildlife, it is a foodplant for caterpillars, bees and birds.

IN BRIEF

🌸	Rubiaceae
🌐	Europe, Asia
🌳	60 x 100cm
🛒	Roadsides, railways, waste ground, fields; can tolerate maritime exposure
☼	Full sun/partial shade/ deep shade
🔧	Grows on most moist, well-drained soils (suffers on very acid soils)
⚘	Hardy perennial
✳	July to August
🌱	Creeping
🍃	Lanceolate
🐝	Flies, beetles
✏	No known hazards
🌿	No known conservation issues

IDENTIFICATION

STEMS Wiry, square upright stems.

LEAVES Star-like whorls of 6–8 thread-like, 1–3cm-long, shiny dark green narrow leaves with a downy underside.

FLOWERS Dense clusters of tiny four-petalled honey-scented hermaphrodite flowers.

SEEDS Two ovate acid-green fruits are formed.

LAUNCH SEEDBOMBS Autumn.

GERMINATION TIME 3–4 months; slow because they need a period of cold to break dormancy.

HARVESTING SEEDS August to September.

PLANT CARE The plant can be divided throughout the season so long as it is kept moist.

CULINARY AND MEDICINAL USES

Edible The leaves are edible and added to salads. • The seed is roasted as a coffee substitute and the flowering tops are distilled to make refreshing drinks. • A yellow dye is obtained from the stem and used as a food colouring. • Flowers are used to coagulate milk.

Medical Lady's bedstraw has a long history of use in herbal medicine though it doesn't feature much as a contemporary medicinal plant. It has been used as a diuretic and to treat skin complaints.

OX-EYE DAISY
Leucanthemum vulgare

Ox-eye daisy is an attractive grassland perennial. It will pretty much grow anywhere except waterlogged sites. The plant has a bitter, pungent juice that deters insect pests. Ox-eye daisy planted by the house is said to repel flies.

IN BRIEF

🌸	Compositae/Asteraceae
🌐	Europe, Asia
🌳	90 x 30cm
🛒	Roadsides, railways, waste ground, riversides, fields
☼	Full sun/partial shade (suffers in deep shade)
🗲	Grows on most moist soils
🌱	Hardy perennial herb
✳	May to September
🌿	Upright, clump-forming
➤	Spatulate
🐝	Bees, flies, beetles, Lepidoptera
✎	No known hazards
🌿	No known conservation issues

IDENTIFICATION

STEMS Stout central stem, occasionally branching, and can be slightly hairy and angular or furrowed.

LEAVES Small clump of basal, dark green, spoon-spatula-shaped, roughly toothed leaves, from which the central stem rises.

FLOWERS Hermaphrodite. Each composite flower is composed of hundreds of tiny flowers or 'florets'.

SEEDS Each floret on a flower head produces an oblongoid dark achene, 1–2mm long with up to ten light furrows. Each flower head produces up to 200 seeds.

LAUNCH SEEDBOMBS Any time. With adequate moisture, seeds will germinate continuously throughout the season.

GERMINATION TIME 1–2 weeks.

HARVESTING SEEDS August to September.

PLANT CARE More dependent on seed for regeneration than on vegetative propagation. • Cut stems down to the ground in winter to promote healthy spring growth. • Generally pest-free but susceptible to stem rot, verticillium wilt and leaf spots.

CULINARY AND MEDICINAL USES

Edible The leaves are eaten in salads in Italy. The root is edible and can be eaten raw. The flower petals can be used as garnishes.
Medical Throughout the ages many medicinal uses have been derived from ox-eye daisy. The whole plant has been used to treat whooping cough, asthma, night sweats, ulcers, conjunctivitis and skin ailments like cuts and bruises.

MARIGOLD
Calendula officinalis

Marigold has been valued for many centuries for its healing powers and is one of the earliest cultivated medicinal flowers. The latter part of its Latin name, 'Officinalis', is the botanical term meaning 'used in the practice of medicine'. Marigold is still a popular garden plant.

IN BRIEF

🌼	Compositae/Asteraceae
🌐	Europe
🌱	30 x 20cm
🛒	Roadsides, railways, waste ground, arable land and cultivated beds
☼	Full sun/partial shade
⛏	Grows on most well-drained moist soils; tolerates poor soils
🌿	Hardy perennial (grown as an annual)
❋	May to October
🌱	Upright
☙	Spatulate or Oblanceolate
🐝	Bees
✎	No known hazards
🌱	No known conservation issues

IDENTIFICATION

STEMS Marigold has stout upright angular branched stems, which are pale green and covered in fine hairs.

LEAVES Alternate light green, covered in fine hairs, with widely spaced teeth.

FLOWERS The bright orange or yellow monoecious daisy-like flowers are borne on a crown-shaped green receptacle. As the flower dies and petals drop off, a circular seed head remains.

SEEDS There is no seed pod; the achene (seeds) are closely curled inwards in the middle of what was the flower head. They are bent/curved, resembling a cat's claws; light brown when dry; spiky, woody and around 5–10mm long.

LAUNCH SEEDBOMBS March to April.

GERMINATION TIME 1–2 weeks.

HARVESTING SEEDS The seeds ripen for collection throughout the growing season from August to November.

PLANT CARE They require very little cultivation apart from the odd thinning out and weeding around them for tidiness.
• To encourage bushiness and more flowers, pinch out the growing tips. • Marigolds will self-seed readily and grow pretty much anywhere. Deadhead to prevent the plant from becoming invasive. • Irrigate regularly during dry periods.

PESTS AND DISEASES
Pests Suffers attacks from slugs and aphids.
Diseases Susceptible to powdery mildew.

CULINARY AND MEDICINAL USES

Edible The flowers and the leaves are edible. • Collect flower heads or petals and dry in a dark place, then seal in an airtight container for later use. They have an interesting flavour – some say sweet and some say bitter! They can be used to flavour and colour fish and meat dishes, soups, salads, custard and baked goods and to colour rice and even cheeses. • The petals can be used in an edible flower salad or as a garnish.

Medical Historically used as a medicinal plant by the Romans and the ancient Greeks, they would drink tea to relieve nervous tension and troubled sleep. • Many herbalists value marigold/calendula for its excellent skin-healing properties; it is aptly called ' the mother of all skin' and many lotions and ointments are made from it for its antiseptic, antibacterial, antiviral, antifungal and anti-inflammatory properties. They are used to treat ailments such as nappy rash, eczema, sunburn, herpes, ulcers, chicken pox, shingles, cuts and grazes and athlete's foot and to soothe irritated nipples for nursing mothers.

OTHER USES The petals can be made into a nourishing skin cream. • A yellow dye is obtained from the flowers to dye fabrics and cosmetics. • A petal infusion can be used to lighten and brighten hair. • The oil has been used in perfumeries.

BELOW MARIGOLD HAS A FANTASTICALLY LONG FLOWERING SEASON WHICH LASTS UNTIL THE EARLY WINTER MONTHS.

FLOWER POWER

Marigold makes an attractive cut flower and also a useful companion plant for the vegetable patch as it helps to deter insect pests.

NASTURTIUM
Tropaeolum majus

This fast-growing frost-tender plant climbs by twisting its leaf stalks around supports and other plants. A good companion plant for radishes, cabbages and fruit trees as it lures pests away from them. It can be container-grown.

IN BRIEF

 Tropaeolaceae

 South America

 3.5 x 1.5m

 Maritime, cultivated beds; has been spotted growing on roadsides, railways, waste ground

 Full sun (cannot grow in shade)

 Drought tolerant; grows on most soils but thrives on free-draining acid soils; tolerates poor soils

 Perennial

 July to September

 Sprawling climber

 Orbicular

 Wind, insects

No known hazards

No known conservation issues

IDENTIFICATION

STEMS Fleshy green trailing stems.

LEAVES Fleshy stalks face upwards towards the light and support the leaves by their centres.

FLOWERS Hermaphrodite, spurred, mildly scented trumpet-shaped flowers in shades of orange, red and yellow.

SEEDS The flower petals wither and clusters of three green seeds form. The seeds look like green curled-up woodlice!

LAUNCH SEEDBOMBS April.

GERMINATION TIME 2 weeks.

HARVESTING SEEDS August to October.

PLANT CARE Provide climbing supports such as a trellis or a wall. • Ensure they are planted in a sunny position and water regularly. • Suffers attacks from slugs, snails, aphids and caterpillars.

CULINARY AND MEDICINAL USES
Edible All parts of the plant are edible, rich in Vitamin C, and have a peppery hot flavour. The blossoms and their foliage can make a colourful garnish for salads and be added to soups and casseroles. The seeds can be pickled like capers.

Medical An infusion of the leaves can be used to increase resistance to bacterial infections and for dermatological and hair conditions. Externally it can be used as an antiseptic wash to treat baldness, minor injuries and skin eruptions.

FEVERFEW
Tanacetum parthenium

This strongly scented short-lived perennial herb was historically grown for its medicinal properties. It spreads rapidly and will grow in some areas where other plants struggle. The dried flower buds can be used as an insecticide.

IDENTIFICATION

STEMS Upright, branching, finely furrowed and hairy stems.

LEAVES Alternate yellow-green leaves, ferny foliage with a citrusy aroma and a downy velvety texture.

FLOWERS The flowers are hermaphrodite, daisy-like with white rays/petals and flat yellow centres and a sweet honey scent.

SEEDS The seed head is a composition of tiny fine-ridged sandy-coloured seeds.

LAUNCH SEEDBOMBS February to March.

GERMINATION TIME 1–2 weeks.

HARVESTING SEEDS Harvest when the petals have dropped and the seed head browns.

PLANT CARE Cutting back to the ground in the autumn improves the shape of the plant. • Self-seeds prolifically; to manage this you may need to deadhead regularly. • Can suffer attacks from aphids, chrysanthemum nematode, leaf miners, snails and slugs.

CULINARY AND MEDICINAL USES

Edible Feverfew has a bitter taste but the leaves and flowers are used as flavouring in savoury pastries, beer and soups.

Medical Feverfew has been used for menstrual pain, migraine headaches, arthritis, fevers and coughs and to aid digestion.

IN BRIEF

	Asteraceae
	South-eastern Europe, Asia
	46 x 45cm
	Roadsides, railways, waste ground, fields, cultivated gardens and walls
	Full sun (suffers in deep shade)
	Drought tolerant; grows on most soils but thrives on free-draining sandy soils (suffers on wet soils)
	Hardy perennial herb
	July to October
	Bushy
	Bipinnatifid
	Bees and flies
	Not recommended during pregnancy or for children under 2 years
	No known conservation issues

BERGAMOT/BEE BALM
Monarda didyma

This clump-forming, easy-to-grow plant, which spreads by running underground stems, has dark green leaves and impressive shaggy scarlet blooms, which will flower continuously throughout the season if deadheaded periodically. It is a good companion plant because it attracts pollinators.

IN BRIEF

- *Lamiaceae*
- North-eastern US
- 90 x 40cm
- Roadsides, railways, waste ground, fields, woodland, cultivated beds
- Partial shade/deep shade
- Grows on most soils but thrives on acid clay soils
- Hardy perennial herb
- June to September
- Clump-forming
- Ovate, spear-shaped
- Bees
- No known hazards (avoid during pregnancy)
- No known conservation issues

IDENTIFICATION

STEMS Square-stemmed, characteristic of the *Lamiaceae* family. They are upright, tough and grooved and covered in fine dense hairs.

LEAVES Leaves are dark green, oval-shaped and coarsely toothed with red leaf veins; they have fine hairs on the underside and are sparsely hairy on the topside. The leaves are aromatic and grow opposite on the stem.

It is noted for attracting wildlife, making it a good companion plant.

ABOVE AS IT DRIES, OPENINGS APPEAR AND THE SEEDS SIMPLY ROLL ONTO THE GROUND.

LEFT ITS ORNAMENTAL QUALITIES ARE ITS CITRUS-SCENTED FLOWERS AND FOLIAGE, MAKING BERGAMOT A GOOD PLANT FOR CUT FLOWERS AND TO ADD TO POTPOURRI.

FLOWERS The hermaphrodite, ragged-looking showy flower heads comprise about 30 long curving tubular flowers, 3–4cm long, above reddish bracts. The flowers come in a range of colours and shades ranging from red and mauve to white.

POLLINATION The tubular shape of bee balm flowers makes it easy for bees to fly in to feed from the sweet nectar and pollinate the plant in the process. • As the seeds ripen the seed pod dries and looks honeycomb-like and button-shaped. The nutlets (seeds) are held at the bottom of the calyx on a kind of pad and when they ripen they simply roll out and onto the ground.

SEEDS The tiny seeds are 1–2mm long, nut-shaped and a light brown colour. It is hard to separate them from the chaff because of their size so don't worry too much.

LAUNCH SEEDBOMBS April to May.

GERMINATION TIME 10–40 days.

HARVESTING SEEDS The seeds heads are ripe and ready to harvest from August to October.

PLANT CARE Cutting back hard after flowering encourages more blooms. • Propagate by division in spring or autumn.

PESTS AND DISEASES
Pests Slugs will attack this plant.
Diseases Bergamot can suffer from mildew if summers are hot and dry.

CULINARY AND MEDICINAL USES
Edible Steep the leaves in water to make a refreshing citrusy tea or add to normal tea to make an Earl Grey tea substitute.
•The young shoot tips, flowers and leaves can be used raw and added to salads as a garnish or as a cake decoration. • The young shoots and leaves can be cooked to enhance the flavour of foods.
Medical Bergamot is frequently used as a domestic medicine for treating digestive disorders and sickness and acts as a carminative for flatulent colic, an expectorant and a diuretic.

OTHER USES An essential oil is obtained and used in perfumery and cosmetic products such as skin and hair treatments. • A soothing ointment, poultice or compress can be made to relieve insect bites and bee stings. • It naturally contains thymol, which has been used as the primary active ingredient in some antiseptic mouthwash formulas.

FOXGLOVE
Digitalis purpurea

The common name foxglove comes from the Anglo-Saxon 'foxes glofa', referring to the tubular flowers suggestive of the gloves of a small animal. Medieval herbalists named the flowers 'witches' thimbles'. Native to Europe, this common biennial thrives in woodland, roadsides and wasteland.

IN BRIEF

🌼	*Scrophulariaceae*
🌐	Europe
🌳	1–2 m x 60cm
🛒	Roadsides, railways, woodland, wasteland, cultivated beds
☼	Sun/partial shade/shade
🔱	Most well-drained soils (prefers acid)
⚘	Biennial
✳	June to September
🌱	Herbaceous biennial
➤	Lanceolate/Ovate
🐝	Bees
✎	All parts of the plant are poisonous
🌿	No threats or conservation issues

IDENTIFICATION

STEMS The foxglove stem is strong, stout and round, woolly and greyish/green.

LEAVES For the first growing season they will develop a rosette of large, oval to lanceolate, alternate, dark green, woolly, toothed leaves that grow at the base of the plant, which will die down over winter and regrow in spring. As the plant continues to develop, the leaves gradually become smaller as they grow up the tall stems.

FLOWERS In the summer of the second growing season a display of tubular bell-shaped purple/pink flowers appear with spotted maroon to purple insides. A spotty landing pad helps the bees identify where to land. The reproductive parts of the flower are positioned in the roof of the flower tunnel and as the bee travels along the tunnel, pollen sticks to its fluffy back.

SEEDS The long oval seed pod contains hundreds of dust-like reddish-brown/black seeds; one plant can produce up to 2 million seeds.

HARVESTING SEEDS Collect seeds June to August when the seed pods begin to brown. It is advisable when collecting foxglove seeds to wear protective gloves to avoid any potential skin irritation and a protective mask so as not to breathe in the fine seeds as they can be an irritant.

LAUNCH SEEDBOMBS June through to the end of August.

GERMINATION TIME 2–3 weeks.

PLANT CARE Use gloves when handling this plant, to avoid any skin irritations. • Because of its tall spikes foxglove may need staking for support if winds are high to avoid wind damage, which could result in broken stems and collapse. • Pinch off faded flowers to prolong their flowering season; when flowering has finished, cut the stems off at the base of the plant. • Foxgloves self-seed readily so you may want to thin out some of the seedlings in your garden. • Water early in the day or late afternoon and feed as necessary.

PESTS AND DISEASES

Pests Foxgloves are relatively pest-free plants – slugs don't like them, so if you have a particularly sluggy garden they are the plant for you!

Diseases Foxgloves need to be planted 45cm apart and provided with adequate drainage and air circulation because they are susceptible to crown rot, leaf spot and powdery mildew.

CULINARY AND MEDICINAL USES

Edible NOT EDIBLE.

Medical Although toxic, historically foxglove was widely used in folk medicine to treat sore throats, as a diuretic and as a compress for bruising and ulcers – it was, however, often fatal.

IT IS STRONGLY RECOMMENDED THAT FOXGLOVE IS NOT USED AS A HOME HERBAL PLANT.

The foxglove is an important foodplant for the humble bumble bee.

ABOVE FOXGLOVES THRIVE IN POOR, WELL-DRAINED SOILS IN DAPPLED SHADE OR FULL SUN AND CAN ADAPT VERY WELL TO ANY GROWING SITUATION.

LEFT WITH THEIR LONG FLOWERING SPIKES, FOXGLOVES PROVIDE STRUCTURE – AND BEAUTY – TO THE BACK OF A GARDEN BORDER.

SWEET CICELY
Myrrhis odorata

Sweet cicely has been cultivated as a medicinal herb for centuries and has been growing wild since the 1770s. It was planted in graveyards in South Wales 'in memory of sweetness' and possibly as a myrrh substitute – it was believed that it bloomed on Christmas Eve.

IN BRIEF

	Umbelliferae/Apiaceae
	Native to the UK, Europe
	200 x 100cm
	Damp ditches on roadsides, railways, waste ground, streamsides, field and woodland margins, hills and mountains
	Full sun/partial shade/deep shade
	Most moist soils; tolerates heavy clay
	Herbaceous perennial herb
	May to June
	Clump-forming
	Tripinnate to Bipinnate
	Bees, beetles and flies
	No known hazards
	No known conservation issues

IDENTIFICATION

STEMS The upright round hollow stems are light green, slightly grooved and hairy, with characteristic leaf sheaths.

LEAVES Aromatic, light green, finely divided, fern-like and up to 50cm long, with slightly downy undersides; often the leaves show characteristic pale patches.

FLOWERS Umbels of flat-topped hermaphrodite clusters of tiny white flowers, each having five unequal, erratically notched white petals and protruding stamens.

SEEDS The seeds are 2cm long, shiny, ribbed and slightly hairy. They resemble seed pods but are actually seeds. • The seeds can be sown green or left to ripen and go black.

LAUNCH SEEDBOMBS July to August if freshly harvested or March to May if purchased.

GERMINATION TIME The seeds need several months of cold weather to germinate.

HARVESTING SEEDS Seeds ripen for harvesting from July to August.

PLANT CARE Sweet cicely is a low-maintenance plant and requires little attention. • If the leaves are required for cooking, it is best to prevent the plant from flowering as the flowering process reduces the flavour of the foliage – but in doing so you will, of course, have no seeds for cooking with! • Sweet cicely self-seeds freely. • Divide the plant from September to May.

PESTS AND DISEASES

Pests Generally pest-free.

Diseases Generally disease-free.

CULINARY AND MEDICINAL USES

Edible Parts of the plant that are edible are the leaves, the root and the flowers. • Leaves can be eaten cooked or raw and have a yummy sweet aniseed flavour. Sweet cicely is one of the dried herbs in a 'bouquet garni' herb mix. A refreshing tea is also made from the leaves. They can be added to stewed gooseberries or rhubarb to reduce the acidity (also, less sugar will be needed). • Roots can be eaten cooked or raw. They can be boiled and used as a root vegetable, or added to salads. The root has a similar flavour to the leaves. • Seeds can be eaten cooked or raw. They can be ground as a spice or chopped and used to flavour salads, cream and bakery goods such as cookies, cakes and fruit pies. Or you could just pop them straight into your mouth like sweets!

Medical Sweet cicely can be used for the treatment of coughs, colds, flatulence and digestive disorders and as a gentle stimulant. • The roots can be made into an ointment, which can treat skin ailments and soothe wounds.

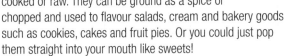

ABOVE THE FLOWERS OPEN EARLY IN THE YEAR AND ARE AMONG THE FIRST FOODPLANTS AVAILABLE FOR NECTAR-LOVING INSECTS, ESPECIALLY BEES.

LEFT SWEET CICELY IS UNIQUE IN THAT IT MANAGES TO SHRUG OFF THE COLD MONTHS AND BEGINS TO PRODUCE FRESH EDIBLE LEAVES IN LATE WINTER.

FERNY FACTS

Historical herbalists said the plant is 'so harmless you cannot use it amiss'.

Its ferny foliage, white flowers and pod-like seeds all have the flavour and aroma of aniseed.

LEMON BALM
Melissa officinalis

Lemon balm is a commonly grown plant used as a household remedy; it has traditionally been made into a tonic to lift the spirits. The Latin name for lemon balm, 'Melissa', is Greek for honey bee; the bees love to feed from the nectar-rich flowers, which appear throughout the summer.

IN BRIEF

	Labiatae
	Europe, Asia, North Africa
	100 x 40cm
	Roadsides, railways, waste ground, cultivated beds
	Full sun/partial shade
	Drought tolerant, grows on most well-drained soils
	Hardy perennial herb
	June to October
	Upright, bushy
	Ovate
	Bees
	Not recommended for those taking thyroid medication
	No known conservation issues

IDENTIFICATION

STEMS Wiry, square-branched upright stems.

LEAVES The green lemon-scented broad leaves are serrated and grow opposite each other up the stem.

FLOWERS The flowers are nectar-full hermaphrodite, 1.3cm long, white tubular in shape with two lips.

SEEDS 1mm long, smooth, black and oval-shaped.

LAUNCH SEEDBOMBS April and September.

GERMINATION TIME 1–3 weeks.

HARVESTING SEEDS The seed heads are ripe for harvest August to October.

PLANT CARE Regularly trimming will promote fresh young leaf growth and prevent vigorous spreading. • To keep foliage flavoursome, cut back some of the flowering stems but leave a few for the bees to enjoy. • Divide in spring or autumn and take cuttings in July/August. It will self-sow readily from seed and can be propagated by taking stem cuttings, which will root easily if placed in a jar of water. • Can be invasive as it grows quickly and spreads easily. A way of tackling this in small cultivated beds is to grow it in a pot and sink the pot into the ground. • Lemon balm requires regular watering but does not require any feeding.

PESTS AND DISEASES
Pests Relatively pest-free.
Diseases Can suffer from verticillium wilt and mint rust.

CULINARY AND MEDICINAL USES

Edible The leaves are the main edible parts used for their lemon aroma and flavour; they can be eaten raw or cooked. • Lemon balm can be used to flavour soups, salads, sauces, poultry stuffings and vegetables, beverages like tea and iced and alcoholic drinks.

Medical Lemon balm has been long used because of its antibacterial and antiviral properties to treat digestive problems, coughs, colds and flu, sleeping problems, menstrual cramps and toothache, for dressing wounds and, according to studies, to calm the nerves and soothe tension. • In the Middles Ages it was used as a cure-all plant to treat skin eruptions and cricked necks and to relieve morning sickness.

OTHER USES Even after harvesting, the leaves and flowers have a long-lasting pleasant aroma and can be used as a potpourri. • The crushed leaves rubbed on the skin can be used as a mosquito repellent.

Lemon balm can also be container-grown outdoors and indoors.

LEMON BALM PESTO

Mix half a cup of Parmesan cheese with 2 cups of lemon balm leaf, 1 cup of walnuts or pine nuts and half a cup of olive oil. Blend and season to taste (garlic can also be added to taste).

ABOVE LEMON BALM IS EASY TO GROW AS A CURE-ALL PLANT AND IS KNOWN FOR ITS FRESH CITRUS FRAGRANT FOLIAGE, WHICH IS USED IN TEAS, SALADS, COLD DRINKS AND DESSERTS.

BORAGE

Borago officinalis

The whole plant is covered in white prickly hairs and has a messy straggling shape. Its star-shaped flowers attract throngs of busy bees. Deters pests and makes an excellent companion plant for tomatoes, strawberries, courgettes and squash.

IN BRIEF

🌸	*Boraginaceae*
🌐	Europe, North Africa
🌲	60 x 80cm
🛒	Disturbed ground, roadsides, railways, waste ground, cultivated beds
☀	Full sun/partial shade
🛠	Grows on most soils (can tolerate nutritionally poor soil)
🌱	Hardy annual
✳	June to October
🌿	Upright, straggling
🍃	Ovate to Lanceolate
🐝	Bees
✏	The leaves not the oil contain small traces of pyrrolizidine alkaloids that may cause liver damage
🌿	No known conservation issues

IDENTIFICATION

STEMS Stout, bristly branched, hollow round stems.

LEAVES The leaves alternate ovate to lanceolate, deep green and wrinkled.

FLOWERS Clusters of hairy flower buds, which open into deep blue star-shaped flowers; the corolla consists of five spreading, purplish lanceolate lobes.

SEEDS The fruit forms throughout the growing season and consists of four black ribbed nutlets.

LAUNCH SEEDBOMBS Any time.

GERMINATION TIME 1–2 weeks.

HARVESTING SEEDS Collect seeds when ripe.

PLANT CARE Borage will seed itself freely and comes up year after year, so it may be prudent to collect some of the seeds before the plant takes over. • It is common for plants to show signs of mildew when grown in dry conditions. • Borage may need protection from wind as it is easily blown over.

CULINARY AND MEDICINAL USES

Edible Borage has a faint cucumber flavour and fragrance. The flowers and leaves are edible (the young leaves are more palatable texture-wise); the flowers are often used as decoration for drinks and salads.

Medical An infusion can treat fevers, chest colds, mouth ulcers, sore throats and menstrual problems. Externally it treats inflammatory swellings, itches, sore eyes and skin conditions.

WILD CHIVES
Allium schoenoprasum

Chives are a common garden plant cultivated for their strongly flavoured edible leaves. They are clump-forming and are ideal for edging beds and paths or letting naturalize randomly. Chives attract wildlife, especially bees and butterflies.

IDENTIFICATION

STEMS Narrow hollow stems sprout from the bulb.

LEAVES Chives have hollow, tubular, bright green leaves with a pleasant onion flavour.

FLOWERS The flowers are hermaphrodite mauve/purple and resemble a pompom. Each flower head is a composition of multiples of tiny flowers.

SEEDS Each tiny flower produces its own papery seed pod, each containing six pointy 2mm black seeds.

LAUNCH SEEDBOMBS March to April.

GERMINATION TIME 1–3 weeks.

HARVESTING SEEDS Harvest seeds when the flowers are spent.

PLANT CARE Regular harvesting gives a continuous supply of young leaves. • To encourage fresh new growth, clumps can be divided in spring. • Relatively pest- and disease-free.

CULINARY AND MEDICINAL USES
Edible The flowers and leaves of chives have a mild onion flavour and can be used for flavouring salads, for Asian dishes, soups, cheeses and dips. • Try adding chives to a salad of home-grown potatoes. Take a pair of scissors and the salad bowl straight to the plant and cut off the leaves/flowers, snipping to the length required in situ.

Medical The whole plant is beneficial for the digestive system and blood circulation.

IN BRIEF

🌸	*Alliaceae*
🌐	Europe, Asia, North America
🌳	45 x 15cm
🛒	Roadsides, railways, waste ground, woodland garden, cultivated beds
☼	Full sun/partial shade
⛏	Most soil types (prefers poor soil); drought tolerant
🌿	Hardy perennial bulb herb
✳	July to August
🌱	Spiky, spreading
🍃	Strap-like, tubular, hollow
🐝	Long-tongued insects like butterflies, moths and bees
✏	No known hazards
🌱	No known conservation issues

WILD MARJORAM
Origanum vulgare

Primarily grown as a pot herb, this woody, sprawling, upright, bushy perennial with aromatic flavoursome foliage is used for culinary and medicinal purposes. The leaves and its tiny summer-borne white/rosy pink flowers are edible. Wild marjoram makes a good companion plant for vegetables and other herbs.

IN BRIEF

🌸	Lamiaceae
🌐	Europe, Asia
🌱	45 x 45cm
🛒	Roadsides, railways, waste ground, arable land, cultivated beds
☼	Full sun (suffers in deep shade)
🗿	Drought tolerant; grows on most soils but thrives on free-draining sandy soils
⁂	Hardy perennial herb
✳	July to September
🌿	Sprawling
🍃	Ovate
🐝	Long-tongued insects like butterflies, moths and bees
✎	Its oil may cause skin irritation. It is not recommended during pregnancy*
🌱	Is not threatened and is a commonly cultivated plant

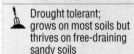

IDENTIFICATION

STEMS Like all plants in the *Lamiaceae* family, wild marjoram has distinctive square stems, which are slightly purple and downy.

LEAVES Its flavoursome aromatic dark green leaves are 2.5cm long, ovate, slightly toothed and borne opposite on the stem.

FLOWERS The dark purple buds at the top of purplish stems open to reveal dense, rose-pink clusters of tiny tubular hermaphrodite flowers.

SEEDS The 0.5mm-long seeds are egg-shaped and rusty in colour, with a smooth testa.

LAUNCH SEEDBOMBS At any time of the year, but the best results are probably achieved from an autumn sowing.

GERMINATION TIME 1–2 weeks.

HARVESTING SEEDS The seeds ripen from August to October and are ready to harvest when the seed heads are dry on the plant.

PLANT CARE Wild marjoram is a great colonizer of sparsely vegetated ground. • Its long roots have numerous root hairs, which enable the absorption of subsoil water in periods of drought. • Can be grown in pots on a sunny patio.

PESTS AND DISEASES

Pests The aromatic quality of wild marjoram deters pests.
Diseases Seldom attacked by diseases, but to prevent risks ensure that air can circulate freely by planting evenly spaced.

CULINARY AND MEDICINAL USES Wild marjoram has been used as a culinary and medicinal herb for thousands of years.

Edible An important herb in Mediterranean cookery. The leaves can be clipped fresh or dried and added to salad dressings, vegetables or chillies. The flowers can be eaten sprinkled over salads or mixed in a potato salad.

BELOW DRIED WILD MARJORAM LEAVES AND FLOWERS CAN BE A TASTY ADDITION TO MANY A MEAL.

• The leaves have a fuller flavour just before the plant flowers and can be used fresh or dried. • Wild marjoram leaves and flowering stems can be made into a refreshing tea when steeped in boiling water for 20 minutes, strained and served with honey to taste.

Medical Wild marjoram is used to promote menstruation and as an antiseptic and has beneficial effects on the digestive and respiratory systems.

• A mild tea made from Wild marjoram can alleviate menstrual pains and promote a restful night's sleep.

NOTE* Wild marjoram has strong sedative effects and should be taken in small doses. It is not recommended as a safe plant to be used during pregnancy.

Attracting much wildlife, it is worthy of inclusion in any herb bed or border.

SEED BREEDING

Wild marjoram is strongly dependent upon seed for reproduction as its vegetative spread is limited. To compensate for this, the plant produces large amounts of seed, which can remain dormant in the soil until the soil has been physically disturbed, especially after fire.

RIGHT MAJORAM'S DISTINCTIVELY SCENTED FOLIAGE TRAVELS FAR ON SUNNY DAYS.

WILD MINT
Mentha arvensis

Wild mint has fresh fragrant foliage (though not as pungent as other mints); the scent becomes more apparent when the leaves are brushed past or bruised. Mint has been cultivated for its culinary and medicinal properties since ancient times, and has been found in Egyptian tombs dating back to 1000 BC.

IN BRIEF

🌸	*Labiatae*
🌐	Europe, Asia
🌳	50 x 100cm
🛒	Roadsides, railways, waste ground, fields, cultivated beds
☀	Full sun/partial shade
🌱	Grows on most moist soils and will grow in heavy clay soil
🌳	Hardy perennial herb
❋	May to October
🌿	Creeping rhizomes
🍃	Ovate, Lanceolate
🐝	Insects, bees
✏	Not recommended during pregnancy
🌿	No known conservation issues

IDENTIFICATION

STEMS Upright, tough, hollow, square, slightly hairy, green-purplish branching stems (adventitious roots may sprout from lower nodes).

LEAVES Opposite, short-stalked, narrowly ovate, sharply toothed, hairy.

FLOWERS The flowers are formed in compact clusters of separate whorls, which are borne in the middle and upper leaf axils. Five united hairy sepals surround the funnel-shaped hermaphrodite flowers. The flowers are pink/purple, 4–7mm long, with five united petals consisting of two upper and three lower lobes; there are two long and three short stamens. The flower matures into a capsule containing four nutlets (seeds).

SEEDS The seeds are tan-coloured, glabrous, oval-shaped and 1mm long. Growing wild mint from seed is very variable and it might be that the flowers and shape are not uniform.

LAUNCH SEEDBOMBS April to October.

GERMINATION TIME 1 week.

HARVESTING SEEDS The seed heads ripen between July and October.

PLANT CARE Wild mint has rather aggressive spreading roots and if you want to prevent them from roaming, restrain them by planting in a container and sink it into the soil. • Division can be easily carried out at almost any time of the year, though there will be more success in spring or autumn. • Cut back in the autumn. • Water regularly.

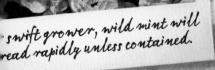

ABOVE THE NECTAR-RICH FLOWERS ARE USED AS A FOODPLANT FOR BEES AND INSECTS.

swift grower, wild mint will spread rapidly unless contained.

PESTS AND DISEASES
Pests Seldom attacked by pests.
Diseases May suffer from mint rust.

CULINARY AND MEDICINAL USES Favoured by the North American tribes, wild mint was used for tea or cold beverages and soups and to flavour pemmican (their native dried meat dish, which was used as an emergency ration).

Edible The leaves of wild mint can be eaten raw or cooked. They have a strong, minty, slightly bitter flavour and are used in salads, cooked foods, desserts, hot and cold beverages and as a garnish. • Add mint to a tasty hedgerow fruit salad using foraged plants such as blackberries and wild strawberries.

Medical Wild mint is used as a home herbal remedy and is valued for its beneficial effects on digestion and its antiseptic, anaesthetic, antispasmodic and aromatic properties. It has agents in it that neutralize inflammation and relieve fevers, colds, headaches, diarrhoea and stomach pains. • Fresh leaves can be chewed and inserted into the nostrils to relieve cold symptoms.

OTHER USES The plant is used as an insect and vermin repellent, and to deodorize houses. • An essential oil is obtained from the plant, which is used in products worldwide.

MINTY FRESH

Wild mint can be grown as a pot herb or planted as a pest deterrent companion for other food plants, such as cabbage, broccoli and tomatoes.

FENNEL
Foeniculum vulgare

Fennel grows naturally over most of Europe; it spread to India and was introduced to the USA in the 1800s. In the UK it grows wild in the south of England and North Wales along the coastal areas. Fennel is widely cultivated for its strong-flavoured leaves and seeds.

IN BRIEF

🌸	*Umbelliferae*
🌐	Europe/naturalized in the UK
🌳	150 x 100cm
🛒	Roadsides, railways, waste ground, fields, maritime, cultivated beds
☀	Full sun (suffers in deep shade)
🔱	Drought tolerant; grows on most fertile moist soils but prefers calcareous soil
🌿	Evergreen perennial herb
❋	August to October
🌾	Upright, feathery
☙	Tripinnate
🐝	Insects
✎	The sap or essential oil has been noted to cause photo-sensitivity
🤲	No known conservation issues

IDENTIFICATION

STEMS Upright, branching, hollow, glaucous-green stems hold alternate leaves and terminate with 5–25cm-wide umbelliferous flower heads.

LEAVES The leaves are composed of long, finely dissected, threadlike, green/blue soft feathery aromatic leaflets.

FLOWERS Large flat umbels of yellow hermaphrodite flowers. The umbels consist of 20 to 50 tiny flowers on short pedicels.

SEEDS 4–10mm-long, grooved, compressed, olive/brown seeds, which are released through openings when the seed pod dries.

LAUNCH SEEDBOMBS March to April.

GERMINATION TIME 2–3 weeks.

HARVESTING SEEDS The seed heads ripen from September to October.

PLANT CARE Fennel tolerates drought but that's not to say it wouldn't like a regular soaking over dry periods. • As it can grow to such a height, fennel will need to be sheltered from strong winds. • Cut back the old growth in winter months. • Fennel can be container-grown in a pot no smaller than 40cm wide and 30cm deep. • To ensure good growth for culinary harvesting, feed every month with a liquid fertilizer.

PESTS AND DISEASES

Pests Can suffer from slug and aphid attacks.

Diseases Generally disease-free.

CULINARY AND MEDICINAL USES Fennel is a herb used worldwide in traditional cooking, often to accompany fish dishes.

Edible The edible parts are the root, stem, leaves and seed. • Fennel is a strong, aromatic and flavoursome herb with culinary and medicinal uses. It can be used to make a herbal tea with a calming and antispasmodic effect, which aids digestive cramps. • Fennel can be found in many forms. The dried seeds have a strong anise flavour; the pollen is the most potent and most expensive form of fennel and is 100 times stronger and sweeter than the seed. It is added as a spice to vegetables and roast meat dishes. The leaves have a delicate flavour and can be added to salads, fish soups and sauces. The bulb is crisp and treated as a root vegetable, which can be eaten raw or sautéed, stewed, grilled or braised. • Fennel is one of the primary ingredients of absinthe.

Medical In India, fennel seeds are eaten raw and are said to improve eyesight. • Fennel is said to improve milk production in nursing mothers and can also be used to treat chronic coughs, colds, flu, bad breath and constipation, and is used as a diuretic.

ABOVE FENNEL IS A FOODPLANT FOR MANY LEPIDOPTERA AND BIRDS.

BELOW DO NOT PLANT NEAR DILL OR CORIANDER AS CROSS-POLLINATION WILL OCCUR, RESULTING IN WEAKER PLANTS.

CUP O' FENNEL

Put a teaspoon of fennel seeds in a teapot and leave to 'steep' for 5 minutes, then strain, pour and enjoy.

BROAD BEAN 'THE SUTTON DWARF'
Vicia faba

As opposed to the common broad bean, this dwarf variety is easier to grow in exposed positions. It's a perfect over-winter crop and the seeds/beans are high in protein and rich in Vitamin C, making it a healthy addition to an edible garden. Like all plants in the *Leguminosae* family, it conditions the soil.

IN BRIEF

- 🌸 *Leguminosae*
- 🌐 North Africa, South-west Asia
- ♣ 45 x 20cm
- 🛒 Cultivated beds
- ☼ Full sun/partial shade, sheltered from winds
- 🌱 Thrives on most well-drained soils (suffers if waterlogged)
- ❋ Hardy annual
- ✳ Spring/summer (depending on when you sow seed)
- 🌿 Bushy, compact
- 🍃 Pinnate
- 🐝 Bees
- ✏ Edible plant
- 🌱 No known conservation issues

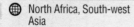

IDENTIFICATION

STEMS Square sectioned hollow stems.

LEAVES The leaves are alternate, divided into 2–7 leaflets and are bluish-grey or green in colour. Unlike most members of the genus, the leaves do not have tendrils for climbing.

FLOWERS Clusters of white-and-black coloured five-petalled flowers.

SEEDS Each cluster develops pods, which contain up to four kidney-shaped light green beans.

LAUNCH SEEDBOMBS February to April or for an early crop October–November (one bean per seedbomb).

GERMINATION TIME 1–2 weeks.

HARVESTING SEEDS When the pod is brown.

PLANT CARE When the plant is spent, cut off the stems and dig the roots back into the soil to make use of captured nitrogen. • The dwarf variety does not require support. • Water regularly. • For winter months the young plants may need protecting with horticultural fleece. • Suffers from aphids.

CULINARY AND MEDICINAL USES
Edible Harvested when the pod is green.

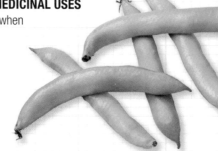

RIGHT BROAD BEAN REMAINS HAVE BEEN FOUND IN ISRAEL DATING BACK TO 6500 BC. IT'S ONE OF THE EARLIEST VEGETABLES TO BE CULTIVATED.

COURGETTE
Cucurbita pepo

Courgettes are simply marrows that are harvested young. They are highly productive, quick and easy to grow vegetables. Courgette seeds are thought to have been brought to the Mediterranean from the Americas during Christopher Columbus' crusades.

IN BRIEF

🌸	*Cucurbitaceae*
🌐	Origins unclear
🌳	30 x 90cm
🛒	Containers, growbags, allotments and cultivated beds
☀	Full sun/partial shade
🔱	Most moist, free-draining soils
🌱	Annual
❄	July to September
🌿	Climber, sprawling
🍃	Deeply cut heart-shaped
🐝	Insects
✏	The sprouting seed produces a toxic substance in its embryo, although it is known as edible
🌱	No known conservation issues

IDENTIFICATION

LEAVES Dull green with a rough texture and can grow bigger than a dinner plate!

FLOWERS The flowers are monoecious. The female flower is yellow and blossoms on the end of each emerging fruit, while the smaller male flower grows from a long stem.

SEEDS Pearly-white flat oval.

LAUNCH SEEDBOMBS May to June.

GERMINATION TIME 1 week.

HARVESTING SEEDS August to October.

PLANT CARE Trailing varieties grow outwards so you may need to train the trails by pinning them to the ground or encouraging them over a support. • Regular harvesting encourages more fruit. • Needs regular watering (especially when in flower). • Feed with an NPK liquid fertilizer once the fruits have started forming. • Apply mulch to prevent water evaporation and water the base of the plant, not the foliage. • Affected by powdery mildew, mosaic virus, aphids and slugs.

CULINARY AND MEDICINAL USES
Edible The watery fruit has a mild flavour and can be used for making bread, cakes, muffins, soups and salads. Stuff and/or batter the flowers. They are a source of vitamins A and C, folate and potassium and are low in calories. • Courgettes freeze well so don't worry if you have too much produce.

SALAD BURNET
Sanguisorba minor

Salad burnet is a clump-forming, low-growing evergreen perennial plant that attracts many different insects into the garden. The plant is self-fertile and tolerates nutritionally poor soil. Can prevent erosion if planted on banks and slopes because its extensive root system clings on to the earth.

IN BRIEF

 Rosaceae

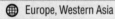 Europe, Western Asia

 60 x 30cm

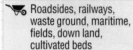 Roadsides, railways, waste ground, maritime, fields, down land, cultivated beds

 Full sun (suffers in shade)

 Drought tolerant; thrives on free-draining, moist, calcareous soils but will tolerate neutral soil

 Hardy perennial herb

 May to August

 Clump-forming

 Pinnate/Lobed

 Wind, bees

No known hazards

No known conservation issues

IDENTIFICATION

STEMS The flowering stem is round, reddish-brown and upright with globular flower heads.

LEAVES Burnet forms rosettes of spiky-toothed greyish leaves and rounded deeply toothed leaflets.

FLOWERS Tiny green flowers arranged within the flower head are female at the upper part with protruding red stigmas, bisexual in the middle, and male at the bottom with pendulous, yellow thread-like stamens.

SEEDS The fruit is a 5mm oblong achene enclosed in a four-winged ridged receptacle. The pale brown seeds are roughly pocked.

LAUNCH SEEDBOMBS Any time of the year (the seeds will germinate in spring or autumn).

GERMINATION TIME 3 weeks.

HARVESTING SEEDS Seeds ripen July to September.

PLANT CARE Burnet is evergreen and will overwinter well. Can be container-grown. • Deadheading will encourage more leafy growth. • Divide in spring. • Burnet will self-sow freely.

CULINARY AND MEDICINAL USES
Edible Burnet smells and tastes of cucumbers and has historically been used to flavour wine, tea and vinegar. It can be added to salads and used as a garnish. • An infusion of the leaves is good for the skin.

RADISH
Raphanus sativus

Thought to have been cultivated for 5,000 years! Radish is a rapidly maturing food crop, primarily grown for its red-skinned rounded/oblong roots. Easy to grow and a good companion plant as it repels pests.

IDENTIFICATION

STEMS Stems rise from the rosette and terminate in flowers.

LEAVES Green/reddish tinged midribs. The lower part of the leaf is deeply lobed with the terminal lobe being larger.

FLOWERS Racemes of hermaphrodite flowers consisting of four pink/light purple petals.

SEEDS Flowers form into a silique containing two or three oval 4–5mm reddish/brown flattish seeds.

LAUNCH SEEDBOMBS From April and in succession every 2–3 weeks until September for a constant supply.

GERMINATION TIME 3–7 days.

HARVESTING SEEDS July to September.

PLANT CARE Provide shade in hot summers and water regularly.
• Food crop radishes are grown fast and picked when young and crisp; if they are left in the ground too long they tend to become fibrous and the roots and leaves become food for various pests.

CULINARY AND MEDICINAL USES

Edible Radish has a peppery flavour and is enjoyed in salads and as a garnish.

Medical The seed can be sprouted and the young seed pods can be eaten raw to stimulate appetite and aid digestion. • Used to treat abdominal bloating, indigestion, asthma, bronchitis and other chest complaints, diarrhoea and as a diuretic. • A radish poultice soothes burns and bruises.

IN BRIEF

🌼	*Brassicacea/Cruciferae*
🌐	Europe, Asia
🌱	40 x 20cm
🛒	Widely cultivated and known to escape to roadsides, railways, waste ground
☼	Full sun/partial shade
🛠	Most moist, free-draining soils (suffers on heavy clay acid soils)
🌿	Annual/biennial
✳	June to August
🥕	Root vegetable
🍃	Oblanceolate/ Pinnate/Lobed
🐝	Bees, flies
✎	No known hazards
🌱	No known conservation issues

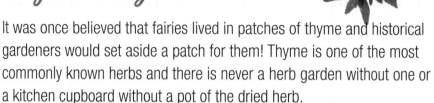

COMMON THYME
Thymus vulgaris

It was once believed that fairies lived in patches of thyme and historical gardeners would set aside a patch for them! Thyme is one of the most commonly known herbs and there is never a herb garden without one or a kitchen cupboard without a pot of the dried herb.

IN BRIEF

🌸	Labiatae
🌐	Europe
🌳	20 x 30cm
🛒	Rocky hillsides, roadsides, railways, waste ground, cultivated beds
☼	Full sun (suffers in the shade)
🔱	Drought tolerant; grows on sandy, alkaline, neutral, free-draining soils
🌱	Evergreen subshrub
✳	June to August
🌿	Woody, ground cover
🍃	Ovate-Lanceolate
🐝	Lepidoptera, bees, flies
✏	Essential oil (thymol) derived from plant is toxic if ingested
🌱	No known conservation issues

IDENTIFICATION

STEMS Gnarled, woody, many-branched stem holding opposite paired leaves.

LEAVES The upper of the leaf is dark green and the underside is grey and downy. The leaf margins are distinctively rolled under.

FLOWERS The tips of the branches hold clustered whorls of hermaphrodite lilac-pinkish-purple tubular two-lipped flowers.

SEEDS Small, globular, and brown/black in colour.

LAUNCH SEEDBOMBS March to April and September to October.

GERMINATION TIME 3–4 weeks.

HARVESTING SEEDS The seed will ripen for harvesting from July to September.

PLANT CARE Thyme can get very woody and the foliage sparse; to prevent this, cut back after flowering. • Leaves can be harvested for fresh use throughout the summer, but the flavour is best just before flowering. • Once plants are established, prune regularly and remove dead flowers and old wood. • Protect from harsh weather conditions, especially wet, by adding a layer of gravel around their bases. • Thyme can be propagated by root cuttings, softwood cuttings, semi-hardwood cuttings, division and layering, as well as by seed. • When you maintain the health of your plant, use it as part of your harvesting routine; the leaves can be used fresh or dried and stored for later use.

PESTS AND DISEASES

Pests Generally pest-free.

Diseases Thyme can get root rot if waterlogged.

CULINARY AND MEDICINAL USES

Edible The leaves can be used fresh or dry – they may be small but they pack a powerful punch! • Thyme goes with whatever you team it with, from stews, soups and sauces to meat dishes, fish dishes and vegetable dishes. A Sunday roast wouldn't be the same without a sprig or two of thyme. • Thyme is an important ingredient of the herb mix 'bouquet garni'.

Medical Thymol is what gives thyme leaves their strong flavour; its medicinal properties are used to treat dental problems and sore gums and it is a main ingredient for mouthwashes. • A herbal tea can be prepared to treat coughs, whooping cough, asthma, colds and flu, as well as the dental problems. Add 2 teaspoons of dried herb per cup of boiling water and steep for 10 minutes.

• An antiseptic essential oil is derived from thyme to treat depression, fatigue, headache, muscular pains and respiratory problems and to make cough medicines. The oil can also be added to a carrier oil (such as almond) to be used as a chest rub, applied to bites and wounds, or added to the bath to soothe arthritic pains.

ABOVE THYME IS NOTED FOR ATTRACTING WILDLIFE, ESPECIALLY EARLY BUTTERFLIES, BEES AND BENEFICIAL FLIES AND WASPS.

The sunnier the spot, the better the flavour!

LEFT THIS HUMBLE LITTLE PLANT CAN TOLERATE POOR SOIL AND IS GREAT FOR GROUND COVER; IT WILL GROW BETWEEN PAVING STONES AND IN CRACKS, WINDOW BOXES AND PATIO POTS, AND MAKES AN IDEAL COMPANION PLANT AS IT FENDS OFF PLANT PREDATORS.

COMMON SORREL
Rumex acetosa

Cultivated as a pot herb in Europe until the Middle Ages but now seen as a wild foodplant noted for attracting wildlife such as Lepidoptera and birds. Because they contain oxalic acid, sorrel leaves have a tangy citrus flavour, making them a popular choice for salads and soups.

IN BRIEF

🌼	*Polygonaceae*
🌐	Europe
🌱	60 x 30cm
🛞	Roadsides, railways, waste land, meadows, cultivated beds, cliffs and coastal dunes
☼	Full sun/partial shade
🪓	Drought tolerant; grows on most moist soils but prefers acid soil
🌲	Perennial herb
✳	May to June
🌿	Spreading
🍃	Oblong heart-shaped
🐝	Wind
✏	Contains oxalic acid, potentially toxic if ingested in large doses
🌿	No known conservation issues

IDENTIFICATION

STEMS Reddish-green, glabrous (hairless), slightly grooved stem with a papery sheath at the base of each leaf.

LEAVES The oblong-heart-shaped, almost acid green leaves have characteristic pointed basal lobes that direct backwards.

FLOWERS Spikes of inconspicuous reddish-green dioecious flower (and later seed) clusters grow above the leaves.

SEEDS Three-sided, pointed at each end, glossy dark brown/red, 3mm long achene (2,000 seeds are produced per plant).

LAUNCH SEEDBOMBS Any time.

GERMINATION TIME 1–2 weeks.

HARVESTING SEEDS June to August.

PLANT CARE For culinary purposes, cut back the flowering stalks in July so the plant can use its energy for leaf production rather than flower production. • Water during the summer months and protect from frosts. • Generally pest- and disease-free.

CULINARY AND MEDICINAL USES

Edible The young leaves and stem are best eaten raw in salads and the older leaves can be used for soups.

Medical Used to treat cancer, sore throats and sinusitis; diuretic and a cooling drink for fevers.

OTHER USES A grey-blue dye can be obtained from the leaves and stems. • Makes an interesting dried flower in arrangements and bouquets.

SPINACH

Spinacia oleracea

Spinach is thought to have originated in ancient Persia and traders introduced it to India, ancient China and then the rest of the world. It's grown for its edible foliage, which can be eaten raw or cooked and is rich in nutrients, especially iron.

IN BRIEF

✿	Amaranthaceae/Chenopodiaceae
🌐	Mediterranean to Central Asia, Afghanistan and India
🌳	30 x 20cm
🛒	Cultivated beds
☼	Full sun/partial shade
⚒	Nutrient-rich moist soil (suffers on acid soils)
🌱	Hardy annual
✳	June to September
🌿	Clump-forming
⬟	Ovate, triangular-based
🐝	Wind
✑	No known hazards
🌿	No known conservation issues

IDENTIFICATION

LEAVES Large, bright green puckered leaves form a rosette at the base of the plant and smaller leaves grow on the flowering stem.

FLOWERS Flower spikes of tiny dioecious greenish-yellow flowers mature into a small fruit cluster.

SEEDS The fruit cluster is tough and dry and contains several black seeds.

LAUNCH SEEDBOMBS March to July.

GERMINATION TIME 1–2 weeks.

HARVESTING SEEDS Harvest the seed July to August when the fruit is dry.

PLANT CARE Pick leaves as required; this also promotes growth of more fresh leaves. • Suffers attacks from aphids and birds. • Can suffer from virus disease, leaf spot and mildew. • To increase crop production, water and feed with an NPK liquid fertilizer on a regular basis.

CULINARY AND MEDICINAL USES

Edible There is a world of ways you can eat spinach – raw in salads, couscous or sandwiches, or cooked. Bear in mind that it shrinks considerably, and don't cook it for more than a minute or two or it will reduce the nutritional value.

OTHER USES Spinach is a good companion for plants like strawberries and cabbages as it deters pests and provides shelter.

COLTSFOOT
Tussilago farfara

Coltsfoot is seen as a common invasive weed which grows abundantly on wasteland and neglected sites by rivers and the seaside. Dandelion-like flowers emerge first and when the seed head forms and the stem dies, hoof-shaped leaves begin to appear.

IN BRIEF

- Asteraceae/Compositae
- Europe, Asia
- 30 x 100cm
- Roadsides, railways, waste ground, maritime, arable fields
- Full sun/partial shade/ deep shade
- Grows on most moist, free-draining soils; thrives on acid soils
- Hardy, rhizomatous herbaceous perennial
- March to April
- Creeping, clump-forming
- Hoof-shaped
- Insects, wind
- The plant contains traces of pyrrolizidine alkaloids, which if ingested in large doses can be potentially toxic to the liver
- No known conservation issues

IDENTIFICATION

STEMS It has upright, unbranched, downy stems, which are covered with alternate reddish/brown scales and terminate in one flower bud.

LEAVES The long-stalked, dark green leaves have grey downy undersides (the young leaves are downy on both sides) and are round-hoof-shaped with slightly toothed edges, divided into five to 12 lobes. The leaves sprout from the rhizome when the flowers and seeds are spent. The leaves can be collected, chopped up and dried between May and July.

FLOWERS Its single-rowed involucre is composed of around 300 yellow strap-like ray florets and around 40 tubular disc florets followed by the seed head, which is a white downy globe.

POLLINATION Visiting insects aid pollination, as does the wind, which blows the pollen from the anthers to the stigma.

SEEDS The seeds are 1mm cylindrical glabrous achenes, which occur in the outer ray florets and, rarely, the inner disc florets. They have a feathery plume to aid wind dispersal and can travel distances of over 4km. • The number of seeds per seed head is around 160.

LAUNCH SEEDBOMBS March to April or September.

GERMINATION TIME 1–2 weeks.

HARVESTING SEEDS Seeds will ripen for harvesting from April to June.

PLANT CARE Coltsfoot can be a problem weed in cultivated gardens; to control spread, cut off flowering heads before they set seed and leaf removal will exhaust the rhizomes. It can be planted in a sunken container to prevent spread in a cultivated bed. • Divide rhizomes any time in the year and plant the divisions straight into their permanent positions.

PESTS AND DISEASES

Pests Slugs attack the flowers and the rhizomes fall prey to wireworms, swift moth larvae and cockchafers.
Diseases Seldom attacked by diseases.

CULINARY AND MEDICINAL USES The flowers can be gathered and dried from March to April and the leaves from May to July or the leaves can be used fresh until autumn.

ABOVE COLTSFOOT CAN GROW COMPETITIVELY AMONG OTHER PLANTS, EVEN WHEN OVERCROWDING OCCURS.

Edible Young flowers and buds have an aniseed flavour and can be eaten raw in salads and as a garnish or cooked in soups; the burnt leaves are used to season foods as a salt substitute. • For tea, pour boiling water over 1.5g of chopped coltsfoot and steep for 20 minutes. Strain before drinking (do not exceed 10g of coltsfoot per day).
Medical Historically used as a relaxant, expectorant, demulcent and diuretic (to relieve cystitis) and to treat lung problems. The dried leaves are smoked to relieve coughs such as whooping cough, bronchitis, etc. • The flowers compressed are a soothing anti-inflammatory, known to relieve joint pain, and can soothe skin conditions such as boils, abscesses and ulcers.
• The plant has been used as a confectionery product to soothe sore throats and chesty coughs.

ABOVE COLTSFOOT IS USED AS A FOODPLANT BY SOME LEPIDOPTERA LARVAE AND BIRDS WILL EVEN TAKE THE WHOLE SEED HEAD.

COLTSFOOT CURER

Coltsfoot has long been used for its medicinal properties, especially for chest and lung conditions, including coughs – 'tussilago' means 'cough suppressant'.

ANISE/ANISEED
Pimpinella anisum

Anise is known as a cultivated crop but it also grows wild. It is in the same family as carrot and parsley. Its aromatic qualities make anise good companion plants and a pest deterrent, while also attracting parasitic wasps to prey on pests.

IN BRIEF

- 🌸 *Umbelliferae*
- 🌐 Middle East, Egypt, Greece, Europe
- 🌳 60 x 40cm
- 🛒 More suited to cultivated beds but can grow in wild situations too
- ☀ Full sun (suffers in deep shade)
- 🌱 Grows in most moist, free-draining soils
- 🌿 Annual herb
- ❋ July to August
- 🍃 Feathery, delicate
- 🍃 LOWER: Heart-shaped; UPPER: Pinnate
- 🐝 Insects
- ✎ Not recommended during pregnancy
- 🌿 No known conservation issues

IDENTIFICATION

STEMS Slender, round, grooved stems branch off at the top.

LEAVES The lower leaves are round to heart-shaped and coarsely toothed, whereas the upper leaves are pinnate, divided, delicately feathery and bright green.

FLOWERS Clusters of tiny yellow/white self-fertile umbelliferae flowers develop.

SEEDS The brown fruit is a downy, flattened ovate and contains two brown ribbed seeds with a distinctive liquorice flavour.

LAUNCH SEEDBOMBS March and September.

GERMINATION TIME Up to 3 weeks.

HARVESTING SEEDS Fruits ripen July to September.

PLANT CARE Protect the plants from winds with support (the seed heads can become quite heavy and pull the plant down). Ensure it is positioned in full sunshine to promote healthy growth.
• Water regularly when hot and during dry weather (always water plants in the morning or the evening).

CULINARY AND MEDICINAL USES
Edible The seeds have a sweet and spicy taste and are used to flavour liqueur and confectioneries.
Medical Anise tea treats coughs, pectoral problems and menstrual cramps; it is carminative, antidepressant, antifungal and also aids digestion.

LIQUORICE
Glycyrrhiza glabra

Liquorice is a robust, woody-stemmed herbaceous perennial with feathery foliage and purplish/blue to white flower spikes. A slow-growing plant that becomes more productive as a root crop after three years of growth.

IDENTIFICATION

LEAVES Feathery leaves with sticky hairs; at night the leaflets droop down.

FLOWERS Hermaphrodite flowers that bloom on long stems followed by smooth-skinned pods containing three brown seeds.

LAUNCH SEEDBOMBS Spring or autumn.

GERMINATION TIME 3–6 weeks (presoaking or scarification may be required for prompt germination).

HARVESTING SEEDS July to August.

PLANT CARE May need protecting from slugs in the first few years of growth • The roots are nitrogen-fixing. • Divide in autumn and plant in new location immediately.

CULINARY AND MEDICINAL USES

Edible Because liquorice is 100 times sweeter than sugar, it is often used in confectionery manufacturing.The roots can be chewed or cut up and used to make herbal tea (1 teaspoon per cup of boiling water).

Medical Cultivated for its roots, liquorice is a widely used plant in Western herbal medicine and has a long history for flavouring and for its medicinal properties, which soothe and treat conditions such as coughs, mouth ulcers, catarrh, bronchitis and sore throats. It is also used for relieving arthritis and for detoxifying and protecting the liver.

IN BRIEF

🌼	Leguminosae
🌐	Europe, Asia
🌳	100 x 100cm
🛒	Roadsides, railways, waste ground, maritime, open sites
☼	Full sun/dappled shade/shade
🛠	Moist, deep, fertile, free-draining soils. (suffers in clay)
🌿	Herbaceous perennial
✳	June to July
🌱	Woody, upright
🍃	Pinnate
🐝	Insects
✏	Should not be used during pregnancy or by people with high blood pressure or kidney disease
🌱	No known conservation issues

SELF-HEAL
Prunella vulgaris

'Self-heal', 'heal-all', 'heart-of-the-earth' – this tiny plant is revered for its powerful properties. It was once believed that self-heal was a holy herb, sent by God to heal all ailments of mankind and animals. Some Native American tribes drank this root tea before hunting to improve their observational powers.

IN BRIEF

🌸	Labiatae
🌐	Europe, Asia, Africa
🌳	20 x 30cm
🛒	Roadsides, railways, waste ground, fields, scrubland, wooded clearings
☀	Full sun/partial shade/shade
🌱	Grows on most moist, well-drained soils
🌿	Hardy perennial herb
✳	May to September
🌿	Upright, groundcover
🍃	Lanceolate
🐝	Bees
✒	No known hazards
🤲	No known conservation issues

IDENTIFICATION

STEMS The stem is square, tough and green with reddish pinstripes running up the stem at the corners of the square, branching at the leaf axis and bearing opposite paired leaves.

LEAVES The leaves are around 10mm long and 5mm wide, lance-shaped and slightly serrated, rich green with reddish tips and short stalks.

FLOWERS The hermaphrodite flowers grow from a purplish/red club-like cluster at the top of the stem. The flowers are deep purple and tubular with two lips, the lower lip being bearded. Directly below the 'flower club' is a pair of stalkless leaves, which act as a kind of collar. After flowering the 'club' becomes a purple-tinged seed head.

SEEDS The seeds are 3mm long, smooth, oval-shaped and rusty brown with slight furrows; when ripe they fall from the seed head and germinate in the ground around the parent plant, but can also be wind-dispersed.

LAUNCH SEEDBOMBS At any time of the year.

GERMINATION TIME Depends on when you sow them, but usually 1–4 weeks.

HARVESTING SEEDS The seeds ripen for harvest from August to September.

PLANT CARE Self-heal can be invasive and self-propagates readily by seed; to keep colonies controlled, deadhead on a regular basis – use this as part of your harvesting regime. • The plant also self-propagates vegetatively through layering (when a stem makes contact with the soil it will develop roots and grow into an individual plant).

PESTS AND DISEASES

Pests Generally pest-free.

Diseases Generally disease-free.

CULINARY AND MEDICINAL USES

Edible The leaves and small flowers of self-heal are edible. The leaves are slightly bitter due to their tannin content. • The flowering tops can be dried in small bunches to be used later; store in a cool, dry, dark place to increase shelf life. • With its slightly herby oregano-like flavour, it can be added to salads and stewed meat dishes. • The whole flower 'clubs' with stem attached would make perfect stirrers for summer drinks, with the bonus of adding a delicate flavour. • For tea to strengthen the immune system, add 28 g/1 oz dried or fresh herb to 600 ml/1 pint of boiling water, steep and add honey to taste. Drink warm or cool in half-cup doses.

Medical Research shows that self-heal can help lower blood pressure and has antibacterial and antibiotic properties. It can be used as a diuretic and to treat mastitis, swollen neck glands, ulcers and conjunctivitis. • Crushed leaves can be used externally as a poultice for wounds, burns, sunburn and bruises.

OTHER USES An olive-green dye is obtained from the flowers and stems, which can be used to dye linen.

ABOVE ONLY BEES CAN POLLINATE SELF-HEAL AS THE CLUSTERED FLOWER TUBES ARE TOO DIFFICULT TO REACH FOR OTHER WILDLIFE.

Purple tubular flowers offer sweet nectar for bees and Lepidoptera.

LEFT SERRATED LEAVES AND UPRIGHT PINSTRIPED SQUARE BRANCHED STEMS HOLD CLUSTERS OF CLUB-LIKE PURPLE FLOWER HEADS.

ECHINACEA (CONEFLOWER)

Echinacea purpurea

Echinacea was used by the Native Americans and the early settlers adopted it as a medicine. It is a fast-growing, versatile herbaceous perennial with a vigorous long flowering season. It requires little attention once established and is an important foodplant for wildlife – particularly birds in the winter.

IN BRIEF

 Compositae/Asteraceae

 USA

 120 x 45cm

 Gravelly hillsides, prairies, open woodland, cultivated beds (has been spotted on roadsides)

 Full sun/partial shade (suffers in deep shade)

 Drought tolerant; grows on most moist, free-draining soils; tolerates clay

 Hardy perennial herb

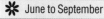 June to September

 Clump-forming, rhizomatous, upright stems

 Lanceolate to Ovate

 Insects

 Rare side effects possible*

No known conservation issues

IDENTIFICATION

STEMS Stoutly round, upright fuzzy stems bear opposite medium/large leaves at the base and intermittently along the stem, almost up to the flower head.

LEAVES Pale to dark green, coarse, toothed and hairy with three prominent veins.

FLOWERS Hermaphrodite, sweet-scented pink/lavender/deep purple daisy-like ray flowers. The petals are 'reflexed' (point downwards) and surround a raised central orange/rich ochre cone.

SEEDS The seed head is dome-shaped, dense and prickly and the seeds are achenes – light brown, 5mm long and cone-shaped with ragged toothed ends.

LAUNCH SEEDBOMBS March to April.

GERMINATION TIME 2–4 weeks.

HARVESTING SEEDS Harvest seeds when the seed heads are ripe and dry – a good clue is when you spot the birds having a munch. Take some seed heads but remember to leave some for the birds as they rely on it in the colder months.

PLANT CARE Because the stems are so tough echinacea doesn't require staking. It copes well with adverse weather conditions. • Cut back stems as the blooms fade to encourage further flower production. • Cut back dead flower stems to the ground in autumn and feed in the spring or autumn. Water regularly but echinacea cannot tolerate being waterlogged. • Divide the plant in spring or autumn and while you are at it you could take some root cuttings.

PESTS AND DISEASES

Pests Can suffer attacks from leaf miners and slugs love it!

Diseases Can be susceptible to powdery mildew and leaf spots.

CULINARY AND MEDICINAL USES Echinacea is commonly used to prevent colds and boost the immune system. It comes in many forms such as tonics, teas, tinctures, tablets, lotions and ointments, root powder, and in dry leaf and flower loose form.

Edible For tea, place 2 teaspoons of dried or fresh leaves in a teapot and cover with 1 cup of boiling water. Leave to steep for 20 minutes, strain and enjoy.

Medical Historically a very important herbal plant which is still widely used to alleviate a variety of ailments such as skin rashes, gynaecological problems, toothache, sore throats, coughs and colds and to boost the immune system; it also helps speed recovery time after an infection.

*USE ECHINACEA WITH CAUTION IF YOU ARE ALLERGIC TO RAGWEED OR PLANTS IN THE *ASTERACEAE/ COMPOSITAE* FAMILY AS IT COULD CAUSE RASHES AND ASTHMA ATTACKS. HOWEVER, SUCH SIDE EFFECTS ARE EXTREMELY RARE, WHICH IS WHY ECHINACEA IS SUCH A WIDELY USED PLANT IN THE WORLD OF HERBAL MEDICINE.

Echinacea is a good summer foodplant for bees and butterflies.

BELOW UPRIGHT STEMS SUPPORT PURPLE DAISY-LIKE FLOWERS, RESEMBLING A FAIRY TUTU, FORMED AROUND ATTRACTIVE ORANGE CENTRAL 'CONES'.

ARNICA

Arnica montana

An upright, elegant, jolly alpine plant. Highly valued since the 16th century for its medicinal and healing properties, which are now widely used in a range of alternative medical practices such as homeopathy and herbalism.

IN BRIEF

🌼	*Asteraceae/Compositae*
🌐	Europe, Asia, North America
🌳	45 x 30cm
🛒	Woodland, hedgerow, pastures, cultivated beds
☀	Full sun/partial shade
🏺	Most moist, free-draining soils; tolerates poor soil
🌱	Herbaceous perennial
❋	June to August
🌿	Alpine plant
🍃	UPPER: Lanceolate; LOWER: Ovate
🐝	Long-tongued insects like butterflies, moths and bees
✏	Poisonous if ingested; should not be used on broken skin
🌿	Scarce in its wild form, possibly due to over-collection as a medicinal herb, and is protected in many parts of Europe

IDENTIFICATION

STEMS The stems are stout, round and downy and sparse in leaf.

LEAVES Bright green toothed leaves; upper leaves are opposite, small and lanceolate, lower more clustered and ovate.

FLOWERS Each stem holds one to three orange/yellow daisy-like hermaphrodite flowers surrounded by soft downy sepals.

SEEDS The seeds are 2mm long dark-grey achene with long bristles (similar to dandelion seed).

LAUNCH SEEDBOMBS Autumn (a period of cold stratification aids germination).

GERMINATION TIME 2–7 weeks.

HARVESTING SEEDS Fruits ripen late summer/early autumn and look like dandelion clocks.

PLANT CARE Shelter from the wind reduces damage and subsequent disease problems like crown rot. • It is unnecessary to feed arnica because its natural environment is usually nutrient deficient. • Divide plants in spring and replant immediately.

CULINARY AND MEDICINAL USES

Edible Arnica is not edible and is poisonous if ingested.
Medical The roots and the dried flowers are used to make ointments and tinctures for the treatment of bruises and sprains, rheumatism, phlebitis and skin inflammations and to soothe common aches and pains.

NOW FOR
THE SEEDS

HOW A SEED IS FORMED

Plants have a symbiotic relationship with wildlife and they rely on visiting insects such as bees and butterflies for pollination. Plants attract pollinators with colourful flowers, which produce nectar in nectaries in the petals' bases. This section will provide a brief explanation of the life cycles of seeds.

REPRODUCTION

The male parts of the flower are called the stamens – there are usually several in each flower. The stamens consist of:

1. The filament, like a stem.
2. The anther rests at the top of the filament and is where the pollen develops.
3. Pollen is released by the anther and is ready to be transferred.

The female part of the flower is called the pistil – it has three sections:

1. The style, a kind of stem, which elevates the stigma so it is reachable.
2. The stigma, which is at the top of the style and has a sticky surface for collecting pollen.
3. The ovary, at the base of the pistil, develops into the fruit, which will contain one or more undeveloped seeds, called ovules.

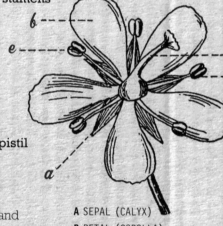

A SEPAL (CALYX)
B PETAL (COROLLA)
C STAMEN
D PISTIL
E ANTHER

POLLINATION PROCESS

A seed cannot form without pollination.
* A grain of pollen lands on the stigma and germinates into a pollen tube.
* The pollen tube grows down the style from the stigma and into the ovary.
* When the pollen tube reaches an ovule, male cells from the pollen grains are released.
* The male cell fuses with a female cell in the ovule.
* Fertilization occurs.

FERTILIZATION

When a flower has been pollinated and fertilization takes place (the fusion of male and female cells), the ovule will mature into a seed up to 25–30 days later.

* The fertilized ovule develops into a seed.
* As the seed grows, the flower changes.
* The corolla withers and dies.
* The ovary wall swells and develops into the fruit. This development determines the dispersal of the seed.

ABOVE AS THE SEED GROWS THE FLOWER WILL BEGIN TO CHANGE AND EVENTUALLY THE PETALS WILL WITHER AND DIE.

GERMINATION

A seed is a plant embryo, an unborn plant with tiny details that show all the characteristics of a complete plant.

During germination, the first thing to appear is the radical, a small root connected to a short stem with the cotyledon(s) (seed leaf or leaves) at the top.

As the root begins to grow a branching system, the stem grows towards the light and forms a hook, which eventually pulls up the cotyledons, usually with the testa (seed coat) still covering them.

A small root connected to a short stem signals the germination process.

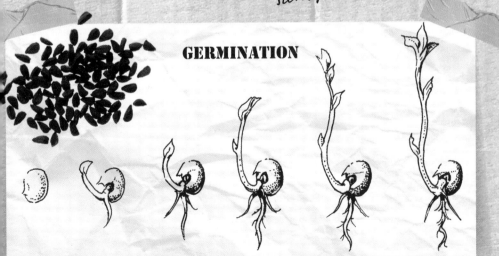

GERMINATION

SEED DISPERSAL

There are many resourceful ways a fruit can form to aid the dispersal of the seed in the most efficient way. The fruit adapts into shapes and sizes that achieve the maximum chances of dispersal and this is dependent on where the plant likes to grow.

A plant that grows by the water, for example, may find water the best way of dispersing seeds and will therefore adapt a fruit that is waterproof and buoyant, such as the coconut. Below are just some ways that a seed can travel some distance from the parent plant to its new growing place.

WIND

Some fruits develop a shape that will enable it to be carried by the wind, such as dandelion parachutes, which have feathery hairs that help them travel long distances.

Maple keys, which have developed propeller-like wings, spin as they fall from the tree and can be taken off by the wind.

And some fruits, such as red campion, have a pepper-pot form that releases the seeds from small holes when the wind blows the stem.

ABOVE THE LIGHT AND FEATHERY HAIRS OF THE DANDELION SEED ALLOW THEM TO TAKE FLIGHT WHEN THE WIND BLOWS.

RIGHT SMALL HOLES AT THE TOP OF THE POPPY HEAD ALLOW SEEDS TO BE DISPERSED WHEN DISTURBED BY WIND OR PERSON.

s can be carried long distances
he flow of rivers or streams.

WATER

Plants that grow by the water rely on water to carry seeds to new locations, such as the palm with its floating coconut fruits or sea kale (*Crambe maritima*). They can travel long distances this way and will either germinate in the water or when they eventually become lodged in a muddy bank.

EXPLOSIVE

As they dry, some fruits will open explosively and expel their seeds – plants from the *Leguminosae* family, for example, like gorse and pea. As the pods are drying, a tension forms in the wall of the pod, which eventually releases like a tight spring and flicks away the seeds.

HITCH-HIKING

Some seeds have developed sticky hooks or spines, which attach to passing animals or humans and hitch a ride for great distances – for example, burdock and our favourite goose grass!

EDIBLE

Yummy fruits are the perfect way to entice animals – they eat the fruit and hidden inside are the seeds. Those seeds which cannot be digested are then expelled as droppings.

Some seeds cannot germinate unless they have been through the digestive system of an animal, as there are acids present that help break down the hard testa.

ABOVE THE HORTICULTURAL HITCH-HIKER: STICKY HOOKS OR SPINES MEAN SOME PLANTS CAN SNEAK A RIDE ON PASSERS-BY AND SPREAD TO NEW AREAS.

Seeds that cannot be digested are dispersed in an animal's droppings.

HARVESTING SEEDS

The best way of building up your seed collection is to harvest them directly from plants. When you collect from the wild you are going straight to the source – it tells you the conditions that plant will thrive in, how big it grows, when it flowers, and when it sets seed.

THE BEST TIME

Harvest seeds when the flower has died and the pod is ripe and swollen. Sometimes the petals are absent or have died and turned brown or black. It varies from plant to plant and happens throughout the summer and into the autumn months.

If you collect the seeds just before they would disperse naturally, you know they will be ripe. Read the plant profiles to find out when the plant sets seed.

THE 'RIGHT' WAY

This varies and is dependent on the type of fruit. With red campion, for example, the seeds are wind-dispersed and the openings appear as the seed pod dries; you can therefore collect the seed just before the openings appear in the pod. This will enable you to collect larger numbers of seed.

HARVESTING KIT AND KABOODLE

If you are a keen collector, take the following whenever you go for a country walk:

* **PAPER BAGS**
* **STRING**
* **SECATEURS OR SCISSORS**
* **PENCIL/PEN**
* **CAMERA**

RIGHT THE RED CAMPION'S STEMS HOLD SEEDPODS AND FLOWERS AT THE SAME TIME, SIMULTANEOUSLY SELF-SEEDING AND ATTRACTING POLLINATORS.

HARVESTING TECHNIQUE

The harvesting technique is simple and applies to most of the plants in this book.

1. **PHOTOGRAPH** the plant you harvest from.

2. **LABEL** each bag even if you don't know what the plant is. Note down distinctive features to help identify your crop when you get home – for example 'blue thistle-like flowers but no spikes on leaves'.

3. **CUT** the seed pod/head off, leaving at least 20cm of stem if possible.

4. **PLACE** in the labelled paper bag with the pods facing down. When you have a good bunch of about 20 stems, tie the bag around the stems.

5. **STORE** the bags suspended anywhere that is dry and warm. I dry them in my kitchen and airing cupboard.

6. **TIME** taken for the pods to be dry enough to release the seeds into the bag also varies from plant to plant; it can take a number of weeks before the seeds are released.

7. **CHECK** every now and then by looking inside the bag to see if the pods have dried. You can often hear them falling.

When you harvest seeds don't be greedy; leave some seed heads behind for the birds to feed on and for future plants to grow.

ABOVE CUT OFF THE POD AND PLACE IN A PAPER BAG, PODS FACING DOWN.

Store your pod-full bags suspended in a warm, dry place.

RIGHT CHECK YOUR BAGS REGULARLY TO SEE WHETHER THE PODS HAVE DRIED – THIS CAN VARY FROM PLANT TO PLANT.

111

SEPARATE THE SEED FROM THE CHAFF

Once your seed pods/seed heads are dry, the seeds are ready to be separated from all the bits of plant debris that surround them – this is called the CHAFF

It can be quite tricky, and being prepared with a clear space like a table or floor and the right tools is a good way to start.

There are a few ways of getting seeds from the pod and it depends on the plant.

Wind-dispersed seeds such as red campion are released from the pods by gently tapping against a bowl or some paper. But some wind-dispersed pods have chambers, which seeds will hide in.

Some pods need to be gently crushed with the end of a rolling pin, with a sieve over a bowl; the sieve will catch the chaff and allow the seed to fall through. Clustered seed heads like marigolds should be pinched apart when dried.

TAKE DOWN your bags plant by plant (so as not to mix up the seeds).
LAY OUT newspaper on the table or floor.
PLACE a sieve over the bowl.
TEAR OPEN the bag, leaving the corners intact as seeds may have collected there.
REMOVE the stems and place them on the newspaper.
TIP the remaining seeds from the bag into the sieve to separate the chaff.
TWEEZER any remaining seeds out of tricky corners.
DRY the seeds for a couple of days in a paper/card container.
PLACE in a labelled airtight jar.
STORE in a cool, dry and dark place. Correct storage can lengthen the lives of seeds considerably.

Once you've collected the dry seeds, place in an airtight container and store in a cool, dry and dark place.

SEEDBOMB
RECIPES

MAKING NATURAL SEEDBOMBS

There are many different recipes for making seedbombs and experimenting is part of the fun! Seedbombs are like miniature gardens – they will be the first soil the seedlings grow in and they need to supply nutrients and have good drainage, like a full-blown garden.

Some people make their own garden compost from household waste such as vegetable peelings and garden trimmings. Others buy it from the local garden centre, or dig earth out of their gardens.

Some seedbomb recipes are simply soggy compost and seeds compressed to make a ball, but these tend to break up in the air or on landing, leaving the seeds much more vulnerable.

It is best to use something to bind the seedbomb and make it hard enough to survive impact with the ground. Whatever you use needs to be water-soluble also, so that water can infiltrate the seedbomb, get to the seeds and break their dormancy.

I have seen some recipes where paper pulp made from egg boxes and office stationery waste has been mixed with compost. As the paper dries, it binds everything together.

Additions such as fertilizers, second-hand tea leaves and coffee grounds provide nutrients to boost the germination process and promote vigorous plant growth.

ABOVE STORE YOUR SEEDS IN AIRTIGHT JARS TO INCREASE THEIR SHELF LIFE.

RIGHT SEEDBOMBS ARE LIKE A HALFWAY HOUSE FOR SEEDS BETWEEN DORMANCY AND THE REAL WORLD.

How much seed you use depends on the size of the seed; for example, the bigger the seed, the more compost and clay you'll need to add to the mixture and the bigger the bomb will need to be in order to accommodate them.

Be generous but not wasteful, because too many seeds will result in overcrowding and bad air circulation, which can make the plants suffer from fungal diseases such as stem rot.

THE SEEDBOMB BASE RECIPE
Ingredients

Makes 6 sizeable seedsbombs

* ✳ **5 TABLESPOONS OF SEED COMPOST**
* ✳ **4 TABLESPOONS OF TERRACOTTA CLAY POWDER**
* ✳ **1 TEASPOON OF SEEDS** (Note: Base this on poppy seeds as a size guide and add half a teaspoon more as the seeds go up in size)
* ✳ **1 TEASPOON OF CHILLI POWDER AS A PEST DETERRENT (OPTIONAL)**
* ✳ **SPRINKLES OF WATER AT INTERVALS** (the geek in me worked out it was about 20ml!)
* ✳ **LIQUID FERTILIZER** if NPK is absent in the compost

TIP: To make larger quantities of seedbomb mixture, use the same proportions but measure using larger containers – mugs rather than tablespoons, for example – and use a bigger bowl, of course!

Mix together well...

YOU WILL NEED

* ✔ **A BOWL**
* ✔ **A STRONG SPOON**
* ✔ **KITCHEN TOWEL OR EGG BOX**
* ✔ **WATER**
* ✔ **A PEN**
* ✔ **YOUR HANDS AND SOME ELBOW GREASE**
* ✔ **AN APRON IF YOU'RE WORRIED ABOUT STAINING YOUR CLOTHES**

Remember when making seedbombs – the bigger the seed, the fewer seeds you need!

AND NOW FOR THE MIXING AND MAKING

1. Pour the compost into your bowl.
2. Pour the clay powder into your bowl.
3. Pour in the seeds.
4. Stir the dry ingredients together until well mixed.
5. Add water in small amounts at a time, mixing and adding until you form a dough-like consistency that sticks together nicely (not too sticky and not too dry).
6. Separate the mixture into six even lumps.
7. Roll each lump into a smooth ball.
8. Place the finished seedbombs on something absorbent like kitchen roll or an egg box.
See **SEEDBOMB Q&A** for any questions.

TIP: When rolling your seedbombs, keep your palms flat to get a rounder shape. If your palms are slightly cupped, you get a shape not unlike a spinning top. Use your fingers to adjust the shape until you are happy with it.

WHAT TO DO NEXT

When you have made your seedbombs, you can:

1. Launch them immediately (if it is the right time of year); they will germinate quicker because they are still moist. Let them dry for a couple of hours so they are not too squidgy and don't lose their shape.
2. Dry them for up to 48 hours. The seeds will remain dormant until activated by water. They can be stored for up to two years and beyond, though some seeds may not germinate if left too long, especially vegetables.

TIP: Dry your seedbombs on a sunny garden wall, shelf, windowsill, radiator or in the airing cupboard.

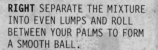

ABOVE POUR YOUR INGREDIENTS INTO A BOWL AND MIX WELL.

ABOVE ADD WATER SLOWLY, MIXING UNTIL YOU HAVE A DOUGH-LIKE CONSISTENCY.

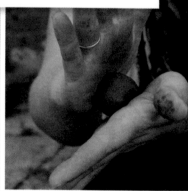

RIGHT SEPARATE THE MIXTURE INTO EVEN LUMPS AND ROLL BETWEEN YOUR PALMS TO FORM A SMOOTH BALL.

Seedbombs for Bees

The plants in this recipe are nectar-rich and a good food source for bees throughout the season. It is important that we grow plants that will provide a source of nectar for the bees, not only to help conserve their colonies but because they are major pollinators for wild flowers and food crops.

Some species of bee are facing extinction and over the last 70 years two species of bee have become nationally wiped out. The Bumblebee Conservation Trust (www.bumblebeeconservation.org) encourages small patches of wild flowers to be planted in field corners, gardens, waste ground, railways, roadside verges and motorway embankments.

And that is where seedbombing is perfect for the job!

PLANTS

FOXGLOVE
Digitalis purpurea

CLOVER
Trifolium pratense

WILD MARJORAM
Oreganum vulgare

CORNFLOWER
Centaurea cyanus

BETONY
Stachys officinalis

LESSER KNAPWEED
Centaurea nigra

Seedbombs for Butterflies

Butterfly Conservation (www.butterfly-conservation.org) believes that butterflies and moths are a fundamental part of our heritage and are indicative of a healthy environment.

It is important to grow food plants right through the season from when they come out of hibernation in spring to autumn, when they need to build up their energy reserves for winter. Grow plants that will provide hibernation, somewhere to lay eggs, food for the larvae (caterpillars) and nectar for the butterfly.

PLANTS

FOXGLOVE
Digitalis purpurea

RED CAMPION
Silene dioica

OX-EYE DAISY
Leucanthemum vulgare

LESSER KNAPWEED
Centaurea nigra

FIELD SCABIOUS
Knautia arvensis

CORN COCKLE
Agrostemma githago

Butterflies are not insects. They are self-propelled flowers.
ROBERT A. HEINLEIN

Seedbombs for Birds

The plants in this recipe attract insect larvae from which the birds will feed, as well as offering a rich seed source throughout the season.

Birds help with seed dispersal and some seeds cannot germinate unless they have first passed through the digestive system of a bird.

The RSPB (www.rspb.org.uk) believes that a healthy bird population is indicative of a healthy planet and the human race depends on this.

Climate change, modern farming methods, road and rail networks, exploitation of our seas and expanding urban areas all pose an enormous threat to birds. We can try to increase the bird population by growing foodplants and creating healthy habitats for them to live and breed in.

PLANTS

SORREL
Rumex acetosa

FIELD SCABIOUS
Knautia arvensis

LESSER KNAPWEED
Centaurea nigra

GREATER HAWKBIT
Leontodon autumnalis

TEASEL
Dispascus fullonum

Seedbombs for the Senses

This recipe has been designed to fill your nostrils with a heavenly sweet scent, and attract and provide food for wildlife too. The plants have been chosen not only for their attractive scents, but for their colour and form.

Perfumes are the feelings of flowers.

HEINRICH HEINE

PLANTS

COWSLIP
Primula veris

FEVERFEW
Tanacetum parthenium

LADY'S BEDSTRAW
Gallium verum

WILD CHAMOMILE
Matricaria recutita

WILD MARJORAM
Origanum vulgare

MEADOWSWEET
Filipendula ulmaria

SWEET CICELY
Myrrhis odorata

Healing Seedbombs

These recipes have plants with healing properties for the mind, body and spirit. The plants will nourish and soothe, relax and enliven, as well as have culinary uses.

Earth laughs in flowers …

RALPH WALDO EMERSON

RIGHT CERTAIN PLANTS SUCH AS WILD MARJORAM PROMOTE RELAXATION AND RESTFULNESS.

WELL-BEING

LEMON BALM
Melissa officinalis

BORAGE
Boragio officinalis

WILD MARJORAM
Oreganum vulgare

WILD CHAMOMILE
Matricaria recutita

LIQUORICE
Glycyrrhiza glabra

WAKE UP

WILD MINT
Mentha arvensis

LEMON BALM
Melissa officinalis

LIQUORICE
Glycyrrhiza glabra

SLEEP WELL

ANISE
Pimpinella anisum

THYME
Thymus vulgaris

MUGWORT
Artemisia vulgaris

COUGHS, COLDS AND HEADACHES

WILD CHAMOMILE
Matricaria recutita

LEMON BALM
Melissa officinalis

WILD MINT
Mentha arvensis

FEVERFEW
Tanacetum parthenium

COLTSFOOT
Tussilago farfara

ECHINACEA
Echinacea purpurea

FENNEL
Foeniculum vulgare

Where flowers bloom,
so does hope.

LADY BIRD JOHNSON

Colourful Seedbombs

The plants in this recipe are perfect for adding a splash of colour to the garden and as a cut flower in the home. These recipes use plants that can be grown in sunny or shaded spots so you can have colour whatever the situation.

Echinacea will flourish in a sunny spot.

PLANTS FOR A SHADY/DAMP SPOT

SELF-HEAL
Prunella vulgaris

BEE BALM
Monarda didyma

BORAGE
Boragio officinalis

PLANTS FOR A SUNNY/DRY SPOT

COMMON POPPY
Papaver rhoeas

ARNICA
Arnica montana

ECHINACEA
Echinacea purpurea

Foody Plants

This section has recipes specifically for culinary uses, like roast dinners, soups and salads. The plants have medicinal uses too, for example to aid digestion or soothe common colds.

To grow a plant, nurture, feed and water it and take it from the land, pop it in a pan and serve it to your family for dinner truly is a wonderful thing.

JOSIE JEFFERY

ABOVE SORREL IS A TASTY AND POPULAR CHOICE TO ADD TO A WILD SALAD.

ALLOTMENT SEEDBOMBS

CHIVES
Allium schoenoprasum

SPINACH
Spinacia oleracea

BROAD BEAN
Vicia faba 'Sutton Dwarf'

COURGETTE
Cucurbita pepo

NASTURTIUM
Tropaeolum majus

WILD SALAD

SORREL
Rumex acetosa edible

MARIGOLD
Calendua officinalis

BORAGE
Borago officinalis

SALAD BURNET
Sanguisorba minor

HOT SALAD

NASTURTIUM
Tropaeolum majus

CHIVES
Allium schoenoprasum

MARIGOLD
Calendua officinalis

RADISH
Raphanus sativus

Seedbomb Q&A

MAKING THE SEEDBOMBS

Q. What do I do if the mixture I make is too sticky to roll into a ball?

A. Either add more earth or leave the mixture to dry for a while.

Q. Should larger seeds be treated differently when making seedbombs?

A. If you are using larger seeds, such as beans, make the earth casing bigger and use fewer seeds.

Q. What do I do if I accidentally forget to put the seeds in before the water?

A. The seeds are usually added to the dry 'mix' because they blend with the earth more evenly. Don't worry, however – just add the seeds and stir in gently but thoroughly.

Q. What happens if the seedbombs go a bit mouldy when they are drying?

A. It could be because your 'drying place' is too humid. Wipe off the mould with a damp cloth and relocate to a drier place, such as above a radiator.

Q. Do the seedbombs have to be dry before you launch them?

A. No. If the time of year is right for the seed to be sown, you can launch them immediately. The moisture content will make germination more rapid.

Q. How do I store the seedbombs?

A. They can be stored when they are completely dry in an airtight container or even paper bags in a dry place for later use.

LAUNCHING THE SEEDBOMBS

Q. How do I know when to launch them?

A. Read the information provided with your seeds.

Q. Why isn't my seedbomb germinating?

A. Maybe it is too dry. A good water will dissolve it a bit and let the light in.

Q. Will all the seeds germinate?

A. It's likely some will germinate this season and some the following seasons.

Q. Do the seedbombs need aftercare once launched?

A. If you can reach them or are growing them as a crop it would be prudent to keep them watered and even give them a liquid feed once in a while. Do some thinning-out if necessary.

ACKNOWLEDGEMENTS

I'd like to thank my folks for my doggedly determined genetics! My siblings for being there for me when I need them, and Steve and my three gorgeous sons for putting up with me! Also my friends for their counsel, and everyone for their love and support.

For their help and advice, I'd like to thank Monica Perdoni at Ivy Press, Janine Nelson at the Museum of Garden History and Brighton and Hove Food Partnership.

I dedicate this book to my little sister Holly, the most beautiful flower in heaven's garden xxxxx

CREDITS

Alamy: 32b, 33t, 34–35b

Catherine McEver, www.stuffyoucanthave.blogspot.com: 27t&m

Corbis: 20t, 25, 101b

Donald Loggins: 34b

Flickr: 33m

Flickr/Ole Peterson: 32t

FLPA: 58t, 77m&b, 101m

Larry Korn, www.onestrawrevolution.net: 14b

PhotoLibrary: 35t

Wayne Blades: 17t

www.suck.uk.com: 26b

GLOSSARY

ACHENE A dry one-seeded fruit, which does not open to release the seed.

ADVENTITIOUS ROOT Root that arises from the stem and not from another root.

ALTERNATE Each leaf grows alternately one at a time along the stem.

ANNUAL Life cycle lasts one season.

ARABLE Land capable of being cultivated.

AXIL Where the leaf joins the stem.

BIENNIAL Completes life cycle in two seasons, germinating and growing the first season and flowering and setting seed in the second.

BIPINNATIFID Pinnate leaves with doubly cut segments.

BRACT A modified leaf protecting the flower.

BULB An underground modified bud and stem used as a food storage organ by dormant plants.

CALCAREOUS Lime-rich.

CHLOROPHYLL Green photosynthetic pigment responsible for trapping radiant light.

CHLOROPLAST The portion of a plant cell that contains chlorophyll.

COMPOUND A flower made up of numerous florets.

CYME Flat-topped cluster of flowers.

DIFFUSION The movement of fluid from an area of higher concentration to an area of lower concentration until a balance is reached.

DIOECIOUS Male and female reproductive organs found on separate plants.

DISC FLORET Small tubular petal-like flower at the centre of the flower head.

ENZYMES Complex chemicals produced by plant cells, which help activate processes such as photosynthesis.

EVERGREEN A plant that retains its leaves all year round.

FLORET Tiny flower.

GERMINATION Transition of seed to seedling.

GLABROUS Smooth, lacking hairs or bristles.

HERBACEOUS Non-woody plants whose leaves and stems die down at the end of the growing season.

HERBICIDE A chemical- or organic-based agent used to kill unwanted plants.

HERMAPHRODITE Male and female organs found on the same flower.

HETEROPHYLLY Plants that have leaves of different shapes on the same plant.

INSECTICIDE Chemical- or organic-based agent used to control damaging insects.

INVOLUCRE A protective whorl of bracts surrounding a flower.

LEPIDOPTERA Moths and butterflies.

LOBED Deeply indented leaves.

MIDRIB A strong central leaf vein.

MONOECIOUS Male and female reproductive organs found on the same plant.

NODE A point on the stem where the leaves emerge.

NODULE An outgrowth from the roots of legumes containing nitrogen-fixing bacteria.

OBLANCEOLATE Lance-shaped but with the

widest part at the tip of
the leaf and the narrowest
at the base.

OBLONG A leaf with a length
greater than the width.

ORBICULAR Circular.

OBLONGOID An elongated
circle.

PANICLE Branched
compound flower of
racemes arranged around
the main floral stem.

PAPPUS A covering of
scales; feathery hairs or
bristles at the apex of
the seed.

PARASITIC WASP Wasp that
feeds on pest insects.

PERENNIAL A plant that lasts
for more than two growing
seasons.

PINNATE A leaf made of
leaflets arranged in a row
on either side of the midrib.

PINNATIPARTITE A leaf
with incisioned lobes
extending over halfway
toward the midrib.

RACEME A cluster of tightly
packed flowers growing in
long thin columns (e.g.
foxglove); the flowers at the
base open first.

RAY FLORET The petal-like
outer floret of a flower
head (e.g. sunflower).

RECEPTACLE Swollen area
at the stem tip where the
flower grows.

REFLEXED Bent downwards
and turned backwards.

RHIZOME An underground
horizontal stem that sends
out roots and shoots.

ROSETTE A low-growing
circular arrangement
of leaves.

RUNNER Horizontal stem
sent out from the base
of a plant, which produces
new plants from buds
along the stem and at
the tips.

SCARIFICATION Cutting the
seed coat to encourage
germination by allowing
water to penetrate the
seed.

SELF-FERTILE Capable of
self-fertilization without
the need for another plant.

SELF-SEED A plant
naturally regenerated
from seed without
human intervention.

SEPALS Modified leaves
that occur outside the
petals and protect the
flower bud.

SILIQUA A long dry seed
capsule with a central
partition to which the
seeds are attached.

STRATIFICATION Pretreating
seeds to help germination
by simulating natural
winter conditions, e.g.
freezing.

SOIL TYPE Classification
of soil based on its sand,
silt, clay and organic
matter content and pH.

SPATULATE Broad and
rounded at the top with
a narrow base.

SPURRED A spiked part
of a flower.

SUBSHRUB A low-growing
woody perennial.

SUBSOIL The soil between
topsoil and bedrock.

TRIFOLIATE A leaf divided
into three leaflets.

TRIPINNATE A leaf divided
three times, as in ferns.

UMBEL A multiple-
stemmed umbrella-
shaped cluster of flowers.

WHORL Arrangement of
leaves, petals, etc. in a
circular or spiral pattern.

VEGETATIVE PROPAGATION
Asexual reproduction
of plants through cuttings,
division, runners.

VIABLE (of seeds) Able to
germinate.

Index